Water Quality - New Perspectives
http://dx.doi.org/10.5772/intechopen.100757
Edited by Sadık Dincer, Hatice Aysun Mercimek Takci and Melis Sumengen Ozdenefe

Assistant to the Editors: Fatıma Masume Uslu and Esra Sündüz Yiğittekin

Contributors
Muthaian Jaya Rajan, Clastin Indira Anish, Ángeles Martinez-Orgániz, Daniel Pineda-Mora, Ricardo Herrera-Navarrete, Viridiana Del Carmen-Niño, Rosa Iris Balbuena Hernández, Job O. Omweno, Albert Getabu, Reuben Omondi, Paul S. Orina, Luís M. Ferreira-Gomes, Ana I.S. Tavares Jorge, Liliana F. Gonçalve, S.M. Soares Almeida, Francisco Riscado dos Santos, Geraldine Seguela, John Richard Littlewood, George Karani, Collins Onyebuchi Okafor, Felicia Ngozi Okoh, Blessing Osose Eromonsele, Ude Ibiam Ude, Lallébila Tampo, Idrissa Kaboré, Seyf-Laye Alfa-Sika Mande, Limam Moctar Bawa, Sadık Dincer, Hatice Aysun Mercimek Takci, Melis Sumengen Ozdenefe, Fikret Buyukkaya Kayis, Aijaz Panhwar, Atta Muhammad, Rashid Abro, Aftab Kandhro, A Rauf Khaskheli, Nusrat Jalbani, Sofia Qaisar, Khadim Gishkori, Veronica A. Okello, Elizabeth N. Ndunda, Abisaki Esitsakha, Mercy Jeptoo

First published in London, United Kingdom, 2024 by IntechOpen
IntechOpen is the global imprint of INTECHOPEN LIMITED, registered in England and Wales, registration number: 11086078, 5 Princes Gate Court, London, SW7 2QJ, United Kingdom

British Library Cataloguing-in-Publication Data
A catalogue record for this book is available from the British Library

Additional hard and PDF copies can be obtained from orders@intechopen.com

Water Quality - New Perspectives
Edited by Sadık Dincer, Hatice Aysun Mercimek Takci and Melis Sumengen Ozdenefe
p. cm.
Print ISBN 978-1-83969-009-9
Online ISBN 978-1-83969-010-5
eBook (PDF) ISBN 978-1-83969-062-4

For EU product safety concerns:
IN TECH d.o.o., Prolaz Marije Krucifikse Kozulić 3, 51000 Rijeka, Croatia,
info@intechopen.com or visit our website at intechopen.com.

Meet the editors

For the past 36 years, Prof. Sadık Dincer has been involved in teaching, research, and academic work at numerous distinguished universities in Turkey. Currently, he is working in the Biology and Biotechnology Departments, Cukurova University, Adana, Turkey. He has published numerous articles in national and international journals manuscripts and book chapters. To date, he has trained twenty-six MSc and eleven Ph.D. students. He received the Technology Development Award from the Scientific and Technological Research Council of Turkey (TÜBİTAK) in 2013 and a national study patent in 2019. His research is focused on bacteriology, microbial ecology, and industrial biotechnology.

Prof. Dr. Hatice Aysun Mercimek Takci received her MSc and Ph.D. in Biotechnology and Biology from Cukurova University, Turkey, in 2007 and 2011, respectively. Since 2009, she has worked at Kilis 7 Aralik University, Turkey. Her teaching interests include microbiology, biotechnology, enzymology, microbial genetics, and bacteriology. She has published fifty-three manuscripts in national and international journals. Her research interests focus on multiple antibiotics and heavy metal resistance in bacteria, the production and characterization of bacterial enzymes, bioremediation by bacteria, and aquatic bacterial contamination. Related to these research areas, she has seventeen projects supported by the Scientific and Technological Research Council of Turkey (TUBITAK) and coordinators of scientific research projects.

Assoc. Prof. Dr. Melis Sumengen Ozdenefe received her BSc, MSc, and Ph.D. in Biology from Çukurova University, Turkey, in 2009, 2011, and 2014, respectively. During her MSc, she spent six months at Anhalt University, Germany, as an international exchange student and a researcher. She has been working in the Department of Biomedical Engineering at Near East University, Northern Cyprus since 2014. Her teaching interests include industrial microbiology, bacteriology, biotechnology, enzymology, and environmental microbiology. Her research areas involve enzymes and biosurfactants produced from various bacteria and fungi for industrial applications, the production and characterization of bacterial enzymes and bacteriocins, the antimicrobial and antioxidant activity of various plant structures, multiple antibiotic resistance, heavy metal resistance in Gram-negative bacteria, and water quality assessment. Dr. Ozdenefe's works have been published in national and international journals, conferences, congresses, and symposiums.

Contents

Preface

Water is the most important resource for the survival of all living organisms and the smooth functioning of ecosystems. Because all forms of life depend on water, a safe and high-quality water supply has become a global concern. Water quality, which is a measure of the suitability of water, varies by location and time. Industrial and agricultural activities, human population growth, and climate change lessen the quality of the Earth's water sources. As the human population increases, so too does the requirement for goods and services. This increasing need for resources has consequently caused the overuse and pollution of water supplies. As such, ensuring water supply and quality requires comprehensive and urgent global action. For efficient management of water resources, it is necessary to know the water quality parameters stated by the World Health Organization (WHO). This book presents a comprehensive overview of water quality along with a series of solutions and recommendations detailing global treatment strategies for water pollution.

Sadık Dincer
Cukurova University,
Adana, Turkey

Hatice Aysun Mercimek Takci
Kilis 7 Aralik University,
Kilis, Turkey

Melis Sumengen Ozdenefe
Near East University,
Nicosia, Turkish Republic of Northern Cyprus

Section 1

Water Treatment Technologies

Chapter 1

Fluoridated Water, Effects and Green Removal Techniques

Veronica A. Okello, Elizabeth N. Ndunda, Abisaki Esitsakha and Mercy Jeptoo

Abstract

Fluoride is a naturally occurring mineral released by rocks into water, soil and air. It elicits dual effects to organisms. Its beneficial effects are effected through water fluoridation to adjust fluoride content in drinking water to acceptable levels that are deemed to prevent tooth decay. Moreover, fluoride itself may be dangerous at high levels. Excessive fluoride causes dental fluorosis and skeletal fluorosis or even severe form of fluorosis known as "crippling skeletal fluorosis," characterized by muscle wasting, immobility and neurological problems. Studies on fluoride levels are important so as to protect organisms from the adverse effect of high fluoride exposure. Various conventional and sensor-based technologies have been applied, leading to the detection of fluoride in water systems across the globe, with some regions reporting levels above the World Health Organization (WHO) limits. This necessitates interventions to reduce the levels of fluoride in drinking water. Green technologies are emerging as viable options for fluoride remediation since they are associated with minimal environmental contamination. Knowledge on fluoride in the environment is a key and therefore, this chapter provides an overview of fluorides, their monitoring in the environment, benefits as well as health effects and removal technologies that range from conventional to green technologies.

Keywords: defluoridation, fluorosis, fluoride in groundwater, drinking water, removal technologies

1. Introduction

The topic on water security and its impact on our daily lives is of utmost concern because many countries are already experiencing water scarcity due to a myriad of reasons. Ironically, water covers about 71% of the earth's surface. Worth noting is that of the world's total surface area, only 1% is covered with fresh water supply present in various natural habitats, such as glaciers, lakes, rivers, groundwater, among others. Of this, only 3% exists as fresh water; 2.5% is unavailable and only 0.5% of the fresh water is available for use by living organisms [1]. The problem of limited supply of fresh water is further exacerbated by the high demand for water due to population growth/explosion, economic growth and changing use patterns, climate change and pollution. The latter can be attributed to primary and/or secondary sources of pollution, which can be from natural and/or manmade sources. These water pollutants include heavy metals, anions, organic matter, persistent organic pollutants (POPs), pathogens and other emerging contaminants such as personal care products (PCPs). According to the World Health Organization (WHO), about 3.4 million people die annually due to water-borne diseases, with reports indicating that over 65 million of the total population in India have been or are affected by fluoridated water [2]. Whereas, the 2018 edition of the United Nations World Water Development Report (UN WWDR) indicated that by 2050, nearly 6 billion people will suffer from clean water scarcity [3].

Of the various pollutants, fluoride is particularly of great concern since it's a naturally available, necessary element for human life but excess intake can lead to adverse health effects [4]. Water fluoridation is a common but debatable practice globally due to the dual nature of fluoride having both positive and negative effects. Undoubtedly, fluoride is an essential trace element that has been used from time immemorial in toothpastes and water as a way to strengthen tooth enamel, however, too much of it can cause harm to the living system. In 2018, 73.0% of the US population on community water systems (207,426,535 people) had access to fluoridated water [5]. Fluoridation involves the addition of fluoride to a calibration amount of 1 mg/L. Some of the salts used in fluoridation include fluorosilicic acid (FSA), silicofluorides (SiFs) and sodium fluorosilicate (NaFSA). Fluoridation of water using sodium fluorosilicic acid and sodium fluorosilicate is reported to accelerate leaching of lead pipes, which increases chances of lead accumulation in children [6].

Studies in several parts of the sub-Saharan Africa (South Africa, Uganda, Kenya, Ethiopia, Sudan, Ghana, Niger, Malawi, Tanzania, Benin and Nigeria, among others) have recorded high fluoride levels in their groundwater [7]. This has been attributed to the geogenic rock structures that are mainly composed of fluorite or fluorspar (CaF_2), topaz ($Al_2SiO_4(F,OH)_2$), apatite ($Ca_5(PO_4)_3(F,OH,Cl)$), biotite ($K(Mg,Fe)_3AlSi_3O_{10}(F,OH)_2$), muscovite ($KAl_2(AlSi_3O_{10})(F,OH)_2$) and cryolite ($Na_3AlF_6$). Fluorite is mainly mined in China, Western Europe and Mexico giving 4 million tons of the annual world production of the fluorite mineral. Very high fluoride levels have particularly been reported in East Africa regions due to the volcanic activities. Several regions in Kenya, specifically the Rift Valley and Central Kenya, have recorded the highest levels of naturally occurring fluoride in the world; up to 23.5 mg/L, which is way above 1.5 mg/L the World Health Organization (WHO) limit [8]. This has been attributed to the leaching of fluoride containing rocks. Sadly, in many residential homes in Kenya, County water supply is not adequate and/or is not available, hence it is supplemented by borehole water which in most cases is contaminated with fluoride ions. This has in turn led to exposure of millions of Kenyans living in these regions to the toxic effects of high fluoride ingestion through the use of untreated groundwater. Fluoride (F^-) is anionic form of fluorine, which is a pale yellow-green univalent poisonous gas, being the 13th most abundant element, constituting about 0.06–0.09% of the earth crust with varied distribution within the environment. Fluorine is a highly reactive element of the halogen family, thus not found in free elemental state but forms organic and inorganic compounds called fluorides. Fluorine is generated through the electrolysis of hydrogen fluoride (HF), finding applications in nuclear power generation by use of uranium hexafluoride (UF6), dielectrics in electric power systems by use of sulfur hexafluoride (SF6), manufacture of fluorinated agents such as chlorine trifluoride (ClF_3) used as an oxidizer, iodine pentafluoride (IF_5) used as a solvent and fluoride salts, such as sodium fluoride, fluorosilicic acid and sodium fluorosilicate used in toothpaste and water fluoridation as well as polytetrafluoroethylene (PTFE) used in nonstick cooking utensils, among many other uses. **Table 1** presents a summary of the chemical properties of fluorine [9].

Fluoride distribution in water, air, soils, rocks and plants is 1.0–38.5 mg/L, 0.1–0.6 μg/L, 150–400 mg/kg, 100–2000 mg/kg and 0.01–42 mg/kg, respectively [10]. Fluoride contamination of the water bodies can occur through both natural and anthropogenic processes. **Figure 1** indicates the distribution of fluoridated water in various regions in the world. The map shows that Africa has a considerable amount of fluoride contamination compared to the rest of the continents in the world. This is followed by Australia and South America (8%), Asia and North America (2%) and Europe (1%). Further reports indicate that 15% of Africa has groundwater, with fluoride concentration exceeding the recommended World Health Organization (WHO) limit of 1.5 mg/L [12].

Groundwater is a major source of water in most developing countries, with over 400 million Africans depending on it for various functions (domestic/industrial). Several water pollution and remediation studies have mainly focused on heavy metals and/or volatile organic pollutants with little emphasis on toxic anions. The study on groundwater fluoridation is of utmost importance because of the dual nature of fluoride. For example, 0.5–1.0 mg/L of fluoride is beneficial for healthy development of bones and teeth, however, a concentration greater than 1.5 mg/L has been linked with several fluorosis and nonfluorosis diseases. A number of side effects dependent on the level of fluoride intoxication have been documented in the literature, broadly divided into osseous tissue and soft tissue effects. These include tooth discoloration, tooth decay, skeletal weakness, neurological problems, liver, kidney and thyroid damage, high blood pressure, acne (fluoroderma), gene and nervous system destruction and seizures. Whereas, liver and kidney are the target organs affected by chronic exposure to fluoride, the anion does not cause seizers but aggravates the effects putting seizure patients at a higher health risk [13]. In a study by Manji et al. (1986), 102 children born and reared in an area of rural Kenya with 2 mg/L fluoride in the drinking water indicated 100% dental fluorosis, 92% of all teeth exhibited a Thylstrup-Fejerskov index (TFI) score of ≥4 and 50% of the children had severe enamel damage [14].

Atomic number	9
Atomic mass	18.9984 g/mol
Electronegativity	4
Density	$1.8*10^{-3}/cm^{-3}$ at 20°C
Melting point	−219.6°C
Boiling point	−188°C
Electronic shell	$[He]2s^22p^5$
van der Waals radius	0.135 nm
Ionic radius	0.136 nm (−); 0.007 (+7)
Isotopes	2
Energy of 1st ionization	1680.6 KJ/mol
Standard potential	−2.87 V

Table 1.
Selected chemical properties of fluorine.

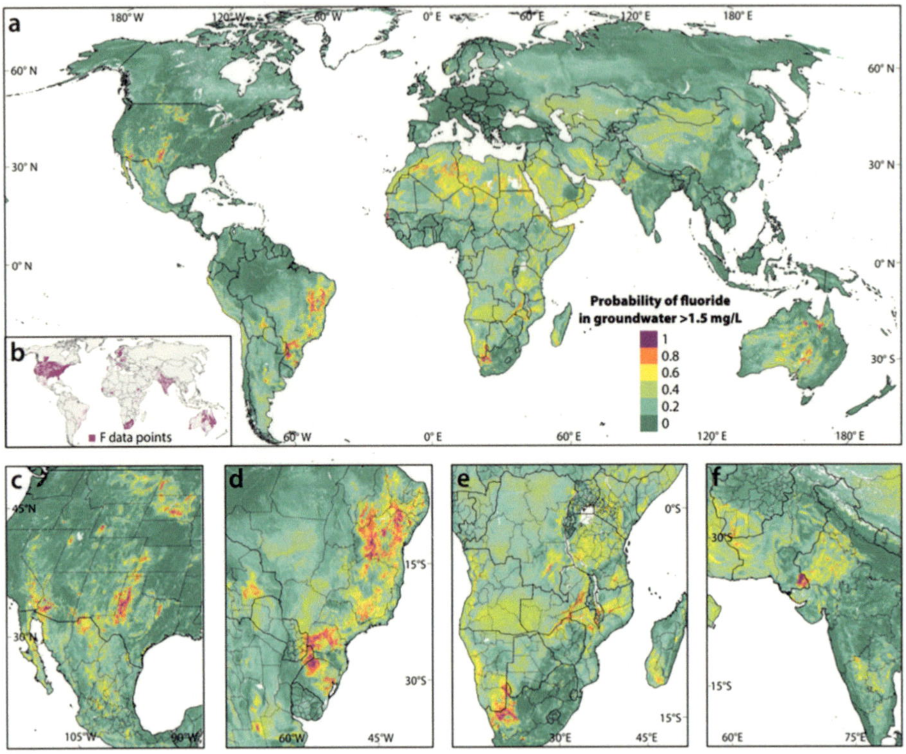

Figure 1.
A probability Map of naturally occurring fluoride in groundwater exceeding the WHO guideline of 1.5 mg/L [11].

This study aims to highlight water fluoridation and distribution, a recent research on the possible negative effects of excess fluoride intake on diverse species, different techniques for the defluoridation of water, such as electro-coagulation, adsorption and membrane processes. Focus will be made on the recent advances of green-based technologies in the treatment of water containing fluorides.

2. A case study of fluoride distribution in south eastern region, Kenya

The south eastern region consisting of Machakos, Makueni and Kitui counties is considered to be arid and semiarid lands (ASALs) and due to the limited amount of rainfall, the region relies on groundwater for irrigation as well as drinking. The region is also reported to contain metamorphic and volcanic rocks that contribute to the groundwater fluoride levels [15]. Drinking water is considered safe for human consumption if the levels of chemical components and physical parameters are within the set standards by the Kenya Bureau of Standards (KEBS) and globally by the WHO. Levels of F^- ranging from 0.6 to 7.17 mg/L have been reported in Makueni County [16] and 4.2 mg/L in Makindu district [17]. Irrigation of crops using this water is a potential pathway for exposure to fluoride and studies have shown that kale and maize pose a health risk to fluoride-related diseases. Concentrations as high as 700 mg/kg in kale and 3.47 mg/kg in soils were reported. The high levels of fluoride in soil and crops were attributed to the presence of apatite, muscovite and biotite, which are fluoride-rich minerals [18].

The earliest reports on fluoride levels in groundwater in Machakos County showed that water in Machakos district was highly contaminated with fluoride with levels up to 16.2 mg L^{-1} [19]. Levels as high as 9.36 mg/L, exceeding the WHO limit of 1.5 mg/L, were reported in borehole water from Mlolongo area [20]. Another recent study in Mumbuni reported fluoride levels at concentrations of 0.252 mg/L and 0.214 for wells and borehole, respectively [21]. From these studies, it is evident that populations in south eastern region are at risk of developing fluoride-related diseases, which is corroborated by a study done in Athi River subcounty, Machakos, where the prevalence of dental fluorosis in children of 12–15 years was reported to be at 93.4% [22]. A recent study by our research group on the quality of borehole, wells and tap water in Machakos County showed that the residents of the area were ingesting fluoride polluted water. Twenty-nine (29) samples were analyzed for their physicochemical properties including fluorides and results showed that all the selected borehole water sources were fluoride-polluted with concentrations ranging from 2 to 7.5 mg/L. These values are very high compared to those of well water and tap water as they also exceeded the acceptable limits for fluoride in drinking water. The high fluoride levels can be explained by the fact that Machakos and its environs have been reported to contain volcanic rocks which could be containing fluoride ions. These fluoride ions get into the waters from weathering of the rocks and through leaching. Only two tap water samples exceeded the acceptable fluoride limits, which could be because the other tap water samples were treated. The fluoride levels of samples collected from selected boreholes, wells and taps in Machakos County are as shown in **Table 2**.

There are few documented studies done in Kitui County to investigate fluoride levels in different water sources, thus calling for further studies to be conducted in the County to conclusively determine the water quality status of the different water sources therein. A physicochemical study done by Nzeve and Matata, 2021 on water samples collected from Kalundu stream and the dam in Kitui County reported fluoride levels between 0.47 mg/L and 1.40 mg/L, which were below the permissible limits set by the WHO [23]. Another water quality assessment undertaken by Wambua et al., 2022 to determine the levels of selected bacteriological and physicochemical parameters in three potable water sources, namely Kiembeni borehole, Mwitasyano River and Kalundu dam, indicated acceptable levels of fluorides ranging from 0.94 to 1.40 mg/L [24]. The lowest concentrations were recorded in the water samples collected from the dams, attributed to the fact that the dam is not static but moves downstream to feed Kalundu River [23]. On the other hand, the highest F^- levels were recorded in the borehole water samples which could have arisen as a result of the leaching of salts from weathering rocks. While the values show that the water is fluoridated, the levels are within acceptable limits and are an indication that the people who consume water from these sources are ingesting fluorides in low concentrations that is beneficial for the prevention of dental caries [25].

3. Determination of fluoride in the environment

Monitoring of pollutants in the environment is important to mitigate the harmful effects of the pollutants to living organisms. Several approaches and techniques exist for the detection and determination of fluorides in water. These include: fluoride ion meter or ion-selective electrodes (ISEs) and ion chromatography (IC) [18, 26], inductively coupled plasma optical emission spectrometry/mass spectrometry (ICP-OES/MS) [27], Fluorine-19 nuclear magnetic resonance spectroscopy (F-NMR) [28], molecular absorption spectrometry (MAS) [29] and gas chromatography-mass spectrometry (GC-MS) [30].

Water samples																	
		Borehole water samples							Well water samples					Tap water samples			
Parameters	Units	BS3	BS4	BS6	BS7	BS8	BS9	BS10	WS1	WS2	WS3	WS4	WS5	TS1	TS2	TS3	TS4	TS5
pH	pH scale	8.05	7.28	7.16	7.41	7.86	8.92		6.26	7.24	7.05	6.84	7.32	6.85	7.38		7.9	8.24
Conductivity	μS/cm	819	638	3626	1572	4919	1036	982	403	3318	928	2387	1799	3294	3294	202	199.9	205
Total hardness	mg/L $CaCO_3$	206	160	880	380	510	ND	ND	66	124	252	260	312	86	86	32	60	40
Chloride	mg/L Cl^-	27.99	33.99	439.8	225.9	644.8	71.48	66.98	21.99	294.9	99.97	279.9	204.9	299.9	299.9	13.99	13.99	13.99
Fluoride	mg/L F^-	**3**	**2.5**	**3.5**	**3**	**6.5**	7.5	**6**	**2.75**	**3.5**	**2**	**2.5**	**2**	5.75	5.75	0.4	0.55	0.75
Sulfate	mg/L SO_4^{2-}	0.96	0.082	1520	48.84	90.57	63.417	46.56	0.237	1252.85	70.57	5.94	46.57	1172.84	1172.84	10.49	5.56	9.48
Nitrates	mg/L NO_3^-	_	25.52	26.4	30.8	44	41.8	9.68	33.44	36.96	88	66	48.4	34.98	34.98	1.892	3.52	2.2

BS = borehole water sample WS = well water sample TS = tap water sample

Table 2.
Physicochemical parameters of water samples collected from different sources in Machakos County in Kenya.

MOFs	LOD	Response range	Selectivity	References
Urea-functionalized MOFs	—	—	$H_2AsO_4^-$ and F^-	[32]
NH_2-MIL-53(Al)	0.31 μmol/L	0.5–100 μmol/L	F^-	[33]
NH_2-MIL-101(Al)	0.05 μmol/L	0.05–8.0 μmol/L	F^-	[34]
NH_2-(UiO-66)	0.229 mg/L	0–50 mg/L	F^-	[35]
Cu(II)-MOF	1.203 ppb	—	F^-	[36]
Y(III)-MOF nanoplates	8.5 ppb	0.05–8.0 mg/L	F^- and pH	[37]

Table 3.
Comparison of various MOFs-based chemosensors for the determination of fluoride ions.

Most of these conventional techniques are not easy to use, are bulky, expensive, require long analysis time and may not be very reliable. The use of chemical sensors, especially functionalized metal organic frameworks (MOFs), has found significant application in the detection of fluorides. This is due to the porous nature and tunability of MOFs with well-positioned ligands that complex with various toxic pollutants. For example, Alhaddad and El-Sheikh, 2021 reported on the use of a salen-cobalt metal organic framework (Co(II)-MOF) to detect fluoride in real water samples [31]. In their study, the photoluminescence spectrum of Co(II)-MOF posted a red shift upon interacting with fluoride ions, even in the presence of other interferents posting limit of detection (LOD) and limit of quantification (LOQ) of 0.24 μg/L and 0.72 μg/L, respectively. Other MOFs-related studies are summarized in **Table 3**.

Chemosensors also suffer certain drawbacks such as the use of supplementary imaging equipment and/or toxic organic solvents. On the other hand, biosensors are more eco-friendly, for example, the use of cell-free biosensor templates has been reported with very low LOD [38]. Toward on-site monitoring of fluoride ions using portable sensors, recently a study by Mukherjee et al., 2020 reported a portable mobile device based on CeO_2@ZrO_2 core-shell nanoparticles' colorimetric detection of F^- in water. The sensor was very sensitive to fluoride in the range of 0.1–5 ppm [39].

4. Benefits and health effects of fluoride

4.1 Benefits of fluoride

Water fluoridation is the adjustment of water fluoride levels to 0.7 mg/L and 1.0 mg/L, with the aim of reducing severity and prevalence of dental caries. Since water intake depends on climatic conditions, as per the WHO guidelines and recommendations in the areas with a warm climate, the optimal fluoride concentration in drinking water should remain below 1 mg/L, while in cooler climates it could go up to 1.2 mg/L [40]. Naturally fluoridated water is beneficial to the human body since fluoride ion is considered an essential element in the development of strong bones, healthy teeth, protection from caries and prevention of osteoporosis [41]. Optimally fluoridated water is deemed practical, economical, effective and safe to prevent dental caries. Fluoride ions present in saliva impede enzymes that breed acid-producing bacteria, hence the tooth enamel is not eroded away [13].

Fluoride ions can also replace the hydroxyl group in hydroxyapatite to form fluoroapatite strengthening the tooth enamel and inhibit the formation of plaque [42]. Children born in areas with fluoridated water derive dual benefits from the water since their salivary glands are exposed to fluoride before tooth eruption hence acting as fluoride reservoirs by continually producing the fluoride and after tooth eruption, fluoridated water ingested is the main source of fluoride reducing tooth decay and promoting strong teeth formation [43]. The use of sodium fluoride (NaF) and sodium monofluorophosphate (Na_2FPO_3) in high doses of 20–30 mg/day to increase trabecular bone density has been used successfully to treat osteoporosis, which is age related. Studies done to determine the prevalence of bone fractures in Chinese populations residing in rural communities of various fluoride concentrations in drinking water reported a decrease in bone fracture which could be associated to an increase in bone mass [44].

4.2 Fluoride toxicity

Climatic conditions, duration of exposure and concentration of fluoride ion determine fluoride toxicity [45]. Studies postulate fluoride concentration of <0.5 mg/L to cause dental caries. Elevated fluoride levels of ≥4 mg/L cause fluorosis. Pearly white flecks on enamel of the teeth surface due to ingestion of fluoridated water are called dental fluorosis. Elevated fluoride levels of 4 mg/L cause dental fluorosis, a defect characterized by lack of mineralization, increased teeth porosity and gap formation. Mild fluorosis is exhibited by white spots on teeth surface, while severe fluorosis is indicated by yellow, brown or black spots [46]. A population estimate shows that 1.7 million people in China and 1 million people in India suffer from dental fluorosis, with recent research suggesting an effect on 70 million people around the globe [41]. Chronic health impact of fluoridated water is indicated by skeletal fluorosis, a condition characterized by increased bone mass and bone density due to overexposure to fluoride concentrations of 5–10 mg/L. Stiffness of bones, joint pains, muscle weakness, periodic pain and chronic fatigue are mild symptoms of the initial stage of skeletal fluorosis. Hardening and stiffening of joints or development of Poker back are clear indications of the intermediate stage, while concentrations >10 mg/L lead to difficulty in walking due to stiff joints and bending bones [8, 47].

Daily consumption of fluoridated groundwater whose fluoride concentration is 3 mg/L leads to the occurrence of skeletal fluorosis in children and adults [47]. In all continents, at least 25 countries have been reported to have endemic fluorosis [48]. Ingestion of fluoride doses over a short period of time leading to poisoning, nausea, abdominal pain, bloody vomiting and diarrhea are effects on the stomach [42]. Collapse with paleness, weakness, shallow breathing, weak heart sounds, wet, cold skin, cyanosis, dilated pupils, hypocalcemia and hyperkalemia are then manifested and in 2 to 4 h death may occur [49]. Other possible effects include muscle paralysis, carpopedal spasms and extremity spasms [13]. Wild herbivores, domestic animal and birds exposed to highly fluoridated water and fluoride in air have also been reported to suffer from dental, nonskeletal and skeletal fluorosis [50]. Plants have the ability to absorb high amounts of fluoride ions from the soil, air, water and store them in the shoot and leaves and have been reported to inhibit photosynthesis and other processes [51]. Consumption of these plants poses a human health risk [50].

4.3 Documented health effects of fluoride in Kenya

In Africa, dental fluorosis has been reported in many countries including South Africa, Tanzania, Uganda, Ethiopia, Kenya, Sudan, Niger, Nigeria, Benin, Ghana and Malawi. On the other hand, few countries, such as Kenya, Senegal, Tanzania and Ethiopia, have reported the prevalence of skeletal fluorosis [7]. Lake Baringo, which is one of the fresh water lakes in Kenya, was found to contain 55 mg/L of fluorides in a research that was conducted in 2021 by the Kenya Marine and Fisheries Research Institute (KEMFRI). This is about 35 times the permissible limit set by WHO. Majority of the residents around the lake who use the water for drinking were reported to complain about back pains, brittle and brown teeth and others showed bowed feet and paralysis. All these are classic symptoms of both dental and skeletal fluorosis, which are primary effects of prolonged ingestion of fluoridated water [52]. A number of studies that have been done on fluorosis in Kenya have shown that 80% of severe forms of fluorosis have been observed in people living along the Kenyan Rift Valley [53]. A study conducted in two health care facilities in Nakuru, a town in the Kenyan Rift Valley, reported the prevalence of dental fluorosis in 86% of the sample population, where 54% showed mild to moderate dental fluorosis and 32% had severe dental fluorosis [25]. This could be inferred to the fact that the area lies along the Rift Valley system, which has been known for volcanic rocks bearing fluoride ions [53].

In a survey conducted by Demarchi et al., 2022 in Nairobi suburbs in Kenya, 80% of the sample population suffered from fluorosis [54]. This could be explained by the fact that the majority of urban population in Kenya are not able to access treated drinking water. As an alternative, most of them depend on groundwater which could be polluted with fluorides. While the belt of the Great Eastern Africa Rift Valley is known for fluoride pollution, other areas outside this region have also been found to be fluoridated. One such area is the Bondo-Rarieda Area in the Kenyan part of the Lake Victoria Basin (LVB). Wambu et al., 2014 found that that 36% of children living in this area, who consume water from ground sources, such as shallow wells, water dams and boreholes from the area, could be at the risk of dental fluorosis [55]. Most of the studies conducted indicated that children are at a higher risk of suffering from fluorosis. In addition, the socioeconomic status is also a factor since it is a determinant of whether or not one is able to access treated drinking water [53].

5. Conventional fluoride removal/remediation techniques

Because of the adverse effects of fluorides, various technologies have been applied to reduce the levels in water. The main methods of defluoridation are adsorption, ion exchange, precipitation, coagulation, membrane processes, distillation and electrolysis. The choice of the method depends on conditions like area, concentration and availability of resources. Among all these techniques, adsorption methods have more advantages because of their greater accessibility, economical use, ease of operation and effectiveness in removing fluoride from water to the maximum extent [56].

5.1 Adsorption-based techniques

Gupta and Ali, 2013 defined adsorption as a stage-by-stage exchange process applied in the removal of substances from fluid phases like gases and liquids [57]. The adsorption process can be operated in physical, chemical and biological systems. Proponents of this technique are of the opinion that in addition to being efficient and economical, it also produces high-quality water [58]. Fluoride removal by adsorption normally follows a three-step process, viz. external mass transfer where the fluoride ions migrate from the bulk solution to the adsorbent surface, adsorption of the fluoride ions onto the adsorbent and finally, the interchanging of fluoride ions that have been adsorbed exchange with the structural elements inside adsorbent particles or their transfer into the inner pores of the adsorbent (intraparticle diffusion) [59], as represented in **Figure 2**.

The most commonly used adsorbents are activated alumina and activated carbon. Adsorption techniques are more applicable for fluoride removal in rural areas because various materials which are good adsorbents can be easily found at low cost. Studies are still underway to design adsorbents from materials that are available in most rural areas, pH independent and less affected by the presence of other ions in water [60, 61]. Discussed are different categories of adsorbent materials.

5.1.1 Carbon-based adsorbents

Carbon-based adsorbents can be used as natural or modified activated carbon containing a complex pore structure that helps increase the surface area available for the adsorption process [62]. Carbon-based materials can be used to prepare activated carbon via chemical or physical activation. Activated carbon can be synthesized from agricultural waste, such as rubber wood, sawdust, coconut shell, sugar beet bagasse, rice straw, bamboo, rattan sawdust and molasses, among others [63]. The synthesis of activated carbon follows several steps including; dehydration and carbonization by heating the carbon material in the absence of oxygen and finally activation by the addition of chemicals [64]. Modified activated carbon has a higher adsorption capacity as compared to unmodified activated carbon adsorbents. Modification of activated carbon is done on the surface of the adsorbents using chemicals by loading different metal hydroxides on activated carbon or treatment with acid to improve its sorption capacity [65]. Activated carbon from plant materials, such as Khat (*Catha edulis*) and *Vitex negundo* plant, has achieved adsorption efficiency of 73 and 89.2% for the removal of fluoride [64, 66]. Also in this category are magnetic carbon-based adsorbents that utilize a magnetic source supported on the surface of carbonaceous materials. These are majorly used for the purification of industrial waste water. These adsorbents have an edge over the other adsorbents, in that they enable the use of alternative and intensified equipment such as the magnetically stabilized fluidized beds, thus, offering new alternatives for water treatment [67].

5.1.2 Calcium-based adsorbents

Calcium-based adsorbents are highly stable and have a high affinity for fluoride ions since calcium binds to fluoride to form new compounds, thus enhancing fluoride removal. Examples of calcium-based adsorbents include the dicalcium phosphate (DCP), which can be prepared by the hydrothermal method and sodium calcium borate glass derived hydroxyapatite (G-HAP) prepared by the immersion of sodium calcium borate glass in 0.1 M dipotassium hydrogen phosphate (K_2HPO_4) solution in the ratio of 50 g/L for 7 days approximately [68]. In a study done by Chen et al. (2022),

Figure 2.
Flowchart summarizing the adsorption process.

calcined eggshells were modified by aging treatment making CaO the main active component. The modified calcined eggshells showed great improvement for fluoride removal by about 29.2% [69]. Modification of biochar from daily manure with calcium enhanced the adsorption efficiency eight times more compared to neat biochar [70]. Bone char is another calcium-based adsorbent that has been widely studied, as it possesses hydroxyapatite (HAp) ($Ca_{10}(PO_4)_6(OH_2)$) with a capacity to remove fluoride through ion exchange as shown in Eq. 1 [71]. Its use though is limited because water treated with animal bones is thought not to be esthetically acceptable [72].

$$Ca_{10}(PO_4)_6(OH)_2 + 2F^- \rightarrow Ca_{10}(PO_4)_6 F_2 + 2OH^- \quad (1)$$

5.1.3 Aluminum-based adsorbents

The most commonly used sorbent is the activated alumina, which is highly porous aluminum oxide with a large surface area. Several factors, such as the characteristics of the feed water, affect the efficiency and effectiveness of active alumina in the removal of fluorides. Active alumina exhibits an increased charge at pH below 6 and is very effective in the removal of fluoride ions. However, its defluoridation effectiveness is diminished at neutral pH [12]. Dehydration of aluminum hydroxide ($Al(OH)_3$) at a temperature range of 300–600°C leads to the formation of a highly porous aluminum oxide (Al_2O_3) with a large surface area. The alumina is greatly preferred for defluoridation due to its discontinuous cationic lattice. Defluoridation by activated alumina is strongly pH dependent, which is a demerit because at pH > 7, silicate and hydroxide become great competitors of fluoride ions and at pH less than 5, activated alumina loses its adsorptivity for it gets dissolved in an acidic environment. Regeneration is required after every 4–5 months and efficiency in fluoride removal decreases after regeneration [73, 74]. **Table 4** shows selected studies that have used alumina in defluoridation.

According to Zhao et al. there are several ways of enhancing the removal of fluorides by an alumina-based adsorbent. The use of sulfuric acid for the activation of alumina-based adsorbents can increase the surface area of alumina and as a result improve their adsorptive characteristics. The use of alumina-based adsorbents has a few disadvantages which include their decreasing efficiency in defluoridation, which is affected by water hardness and their solubility in high fluoride concentrations due to the formation of monomeric aluminum fluoride and aluminum hydroxyl fluoride complexes [59].

5.2 Coagulation-precipitation (Nalgonda technique)

Nalgonda technique involves the addition of aluminum salt, lime and bleaching powder in sequence followed by rapid mixing, then coagulation, sedimentation, filtration and disinfection. Nalgonda technique utilizes alum as a coagulant with lime and bleaching powder as disinfectants, hence it is simple and economical for fluoride removal from drinking water. The optimum pH range for maximum removal of fluoride is reported to be between 5.5 and 7.5. Nalgonda technique was used extensively in Nalgonda village in India under the Rajiv Gandhi National Drinking Water Mission (RGNDWM), though it has limitations. This method removes approximately 18–33% of fluoride through floccule formation while about 67–82% remains as the soluble fluoroaluminate (AlFX) complex ion which results in aluminum toxicity in drinking

Type of alumina	Contact time (min)	pH	Efficiency (%)	References
Manganese dioxide coated	180	5.5	98	[75]
Mesoporous aluminum oxide loaded on calcium oxide	15	3–11.5	90	[76]
Alum and lime	240	5.5–7.5	18–33	[13]
Magnesia/alumina	140	6.3–7.3	85	[77]
Magnesia amended activated alumina	180		95	[78]
Lanthanum oxide impregnated granular activated alumina	30–480	3.9–9.6	70.5–77.2	[79]
Alumina modified expanded graphite	120	3–7	94.4	[80]
Red mud	20	5–7	70–80	[81]
Lanthanum impregnated bauxite	20–250	5–7	99	[82]
Bauxite	15–120	5–6	94	[82]
Pyrophyllite	20	2.8–4.9	85	[83]

Table 4.
Use of alumina in defluoridation and percentage efficiency.

water due to a rise in the permissible Al limit of 0.2 mg Al/L [13]. Sludge disposal is another environmental health problem associated with Nalgonda technique of defluoridation [84].

5.3 Ion exchange

Ion exchange involves the removal of fluoride in water through replacement of ions that are held loosely by ion exchange resins with fluoride ion. There are two types of ion exchange resins according to the functional group that is attached to the polymer matrix. Anion exchange resins exchange negatively charged ions (like fluoride), whereas cation exchange resins exchange positively charged ions from the solution [85]. These resins have small porous beads and are insoluble in most organic solvents and water, making them suitable for fluoride removal. The ion exchange process of fluoride removal involves passing water through the bed of ion exchange resins that are usually polymeric. Polystyrene anion exchange resin and basic quaternary ammonium type are used to remove fluoride according to the reaction 2:

$$Res - NR^{+3}\, Cl^{-} + \mathrm{F}^{-} \rightarrow Res - NR^{+3}\, \mathrm{F}^{-} + Cl^{-} \qquad (2)$$

The fluoride ions substitute the chloride ions of the resin. When all the sites of the resin are fully replaced by fluoride ions, it is regenerated by passing brine solution through the resin. Chloride ions then substitute the fluoride ions recharging the resin, hence it can be reused. Selective defluoridation with an efficiency of 90–95% has been reported using an ion exchange cyclic process. Though ion exchange techniques are excellent in defluoridation, the resin and its maintenance are costly, whereas the treated water has a high chloride concentration with low pH [45]. Modifications of the polymeric resin to increase fluoride removal efficiency include impregnation with various metals, viz. zirconium, titanium and iron, among others. Anion exchange resin impregnated with zirconium reported increased fluoride removal at 60% efficiency [72].

5.4 Membrane technologies

Commonly used membrane-based technologies that include ultrafiltration (UF), reverse osmosis (RO) and nanofiltration (NF) are among the best available defluoridation technologies [85]. Fluoride removal efficiency using reverse osmosis is more than 99%. The major setback in using reverse osmosis is that the technique is uneconomical for average income family and rural populations because it requires special equipment, specialized training for operators and electricity for its operation. Reverse osmosis is therefore normally used in rural sectors of the developing countries where energy and competent human resource are often unavailable [45]. However, membrane technologies can run on renewable energy sources like photovoltaic or wind turbines, which are helpful in reducing waste generation and minimizing carbon emissions to the atmosphere [86]. The other disadvantage of RO membranes is that they can discard a considerable amount of the feed water as a reject stream [87] since they reject ions based on their size and electrical charge [88].

6. Emerging technologies and green defluoridation techniques

6.1 Emerging technologies

There are quite a number of new emerging technologies reported for defluoridation that involve distillation, precipitation and a combination of principles. The Crystalactor® is one of the emerging technologies developed by a Dutch company that consists of a pellet reactor filled with suitable seed material to produce pellets of the target material through crystallization. Defluoridation is achieved by the formation of reusable, extremely low water content and highly purified calcium fluoride pellets. The Crystalactor® is a compact and low-cost technology compared to conventional precipitation techniques. The technology is recommended for the treatment of high fluoride waters (>10 mg/L) and to attain concentrations below 1 mg/L, a second treatment is often required [49]. The Memstill® technology is another technology that is a membrane-based distillation idea developed by the Netherlands Organization for Applied Scientific Research (TNO). This technology improves ecology and economy of the existing desalination technologies for brackish and sea water. The technology successfully removes other anions such as fluoride and arsenic. Memstill® technology combines multistage flush and multi-effect distillation modes using one membrane. In this technology, cold feed water takes up heat in the condenser channel through condensation of water vapor, then a small amount of (waste) heat is added, and flows countercurrently back via the membrane channel. The added heat evaporates water through the membrane which is discharged as cold condensate. The cooled brine is extra concentrated in a next module or disposed. The Memstill® technology is cheaper compared to reverse osmosis and distillation. Memstill® technology for small-scale applications using solar heat is yet to be achieved [89].

The water pyramid is employed in tropical rural areas. It uses solar energy to produce clean drinking water from saline, brackish or polluted water [13, 89].

The technology also removes fluoride by use of a water pyramid with a total area of 600 m^2 placed under favorable tropical conditions, to produce about 1250 liters of fresh water per day. The rate of production depends on climate, temperature, cloud cover and wind activity. Solar energy drives the desalination, while the energy required for pressuring the Water Pyramid® is obtained using solar cells combined with a battery backup system. Intermittent peak demands in electricity are accomplished using a small generator. The use of a porous membrane leads to purification of water using solar energy similar to the water pyramid. In this technique, humidity in the evaporation chamber is increased by water which sweats through the membrane and evaporates on the membrane surface. Temperature difference leads to condensation of pure water on the cooler surface of the system [49]. Intensity of solar radiation determines the quantity of water. Brine has to be drained periodically to avoid crystallization. Seawater and water contaminated with heavy metals, oil residue, boron and fluoride can be purified if the pH range is 5–11.

Boiling with brushite and calcite is another technology where fluoridated water is boiled with brushite ($CaHPO_4.2H_2O$) and calcite ($CaCO_3$) to form fluoroapatite, which results in defluoridation [49]. *In situ* treatment methods for fluoride removal rely on the dilution of fluoride concentration of groundwater. They include check dams, percolation tanks and recharge pots. Groundwater from wells closer to the check dams, percolation tanks and recharge pots have their water fluoride levels reduced through recharge, hence their quality improves compared to water in wells faraway [68]. These methods mainly rely on recharge using rainwater, hence may be limiting in arid and semiarid areas which have little to no rainfall.

6.2 Green technologies for defluoridation

Whereas, remediation technologies have continued to evolve, green remediation strategies based on the 12 principles of green chemistry have taken a center stage since they offer significant potential for increasing the net benefit of the detoxification process by cutting costs, yielding greater efficiencies at a reduced and/or no negative environmental impact. Technologies based on biological remediation capability of plants and microorganisms meet this need and play a vital role in remediation of pollutants. Biological methods, although require more time, are far cheaper than the physical- and chemical-based strategies [90]. Thus, efficient, applicable, low-cost and sustainable water treatment technologies are required to keep up with the surging demands of clean water supplies.

Green remediation techniques have become more significant in the recent past due to less environmental impact, low cost and higher social appeal as compared to other chemical-based conventional techniques. Moreover, the integration of several green remediation techniques may have a higher synergistic effect on remediation efficiency. **Figure 3** indicates the growing number of publications in the application of eco-friendly remediation processes of pollutants from water, and defluoridation methodologies as reported in the Chemical Abstracts Service (CAS) SciFinder search done on May 08, 2023 for the last 22 years. A search done under the keywords, "green and sustainable chemistr" and "green remediation of fluorides in water" and fluoridation studies, indicated 38,284, 341,241 and 8783 related publications, respectively. From these statistics, it is clear that the number of fluoride-related studies and green remediation studies has increased tremendously over the years.

Depending on the type of adsorbent used, adsorption would qualify as a green remediation technique. Adsorption is a viable remediation technique with a myriad

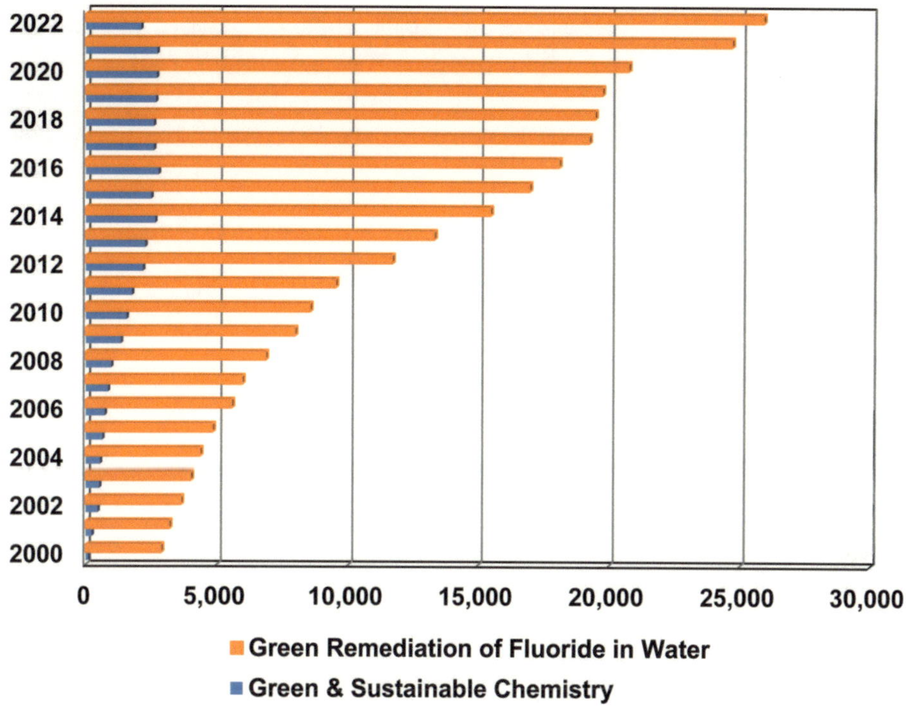

Figure 3.
Yearly statistics on studies related to green chemistry and fluoridation.

of advantages, viz. cost, effectiveness, higher accessibility, simplicity of design, large number of adsorbents, local availability and cheapness. However, it poses various disadvantages such as pH adjustment and decreased adsorption capacity with repeated use of the regenerated sorbent. Adsorption techniques are by far the most commonly used in the developing world due to their simplicity in application and low cost. Adsorbents can be derived from various naturally occurring materials including indigenous minerals and plants, however, they can also be synthetically derived. Effects of major adsorption parameters, viz. pH, dose of adsorbent, rate of stirring, contact time and initial concentration of adsorbate on pollutant removal, must be studied to determine the optimum sorption conditions. As indicated earlier, adsorbents can be natural-based (e.g., red mud, chitosan, bauxite, soil, leaves and bark of trees), carbon-based (e.g., activated carbon), iron-based (e.g., iron oxide-hydroxide nanoparticles, calcium-based (e.g., crushed limestone) and/or alumina/aluminum-based (e.g., gibbsite containing materials). **Table 5** presents a summary of selected naturally available adsorbents and their respective optimum sorbent doses for removal of fluoride ions in water. The majority of these studies were reported to follow first-order rate mechanism and Freundlich or Langmuir isotherms.

Obijole et al. studied aluminosilicate activated clay hydrothermally treated for fluoride and pathogen removal from water at pH 5.8. The results indicated a maximum adsorption capacity of 1.75 mg/g with a 53% fluoride removal at 25 °C [102]. The study by Cherukumilli et al. investigated the use of minimally processed (dried/milled) bauxite ore as an adsorbent for remediating fluoride-contaminated groundwater with doses of ~10–23 g/L effectively remediating 10 mg of fluoride ions per liter of water [103]. Ayoob et al. reported on carbonized form of the biomass of water hyacinth (*Eichhornia crassipes*), after thermal activation at 600 °C, and observed a

Adsorbent	Optimum sorbent dose	Optimum pH range	Removal efficiency (%)	References
Brick powder	0.2–2 g/100 mL	6.0–8.0	54.4	[91]
	0.2–2 g/100 mL	8.0	56.8	[92]
Activated bagasse	4 g/L	6.0	56.4	[61]
Sawdust raw			49.8	
Wheat straw raw			40.2	
Hydrated cement	1 g	8.2	91	[93]
	10 g/L	3–10	92.37	[94]
Mechanochemically activated kaolinites	2.5 g	3	90	[95]
Tamarind seeds	2 g/L	7.0	100	[96]
Blue-green algae, *Phormidium sp.*	4.5 g	3.5–4.5	60.0	[97]
Banana peel	2 g	5	86.5	[98]
Tea leaves loaded with Al/Fe oxides	2 g/L	4.0–8.0	85	[99]
Neem leaves powder	5.0 g/L	5.0–7.0	80	[60]
Moringa oleifera seed's powder	2.5 g/L	6–7	92.3	[100]
Maerua subcordata root powder	200 mg/L	—	66.2	[101]

Table 5.
List of some natural adsorbents and optimum conditions for the removal of fluoride ions.

removal capacity of 4.4 mg/g [57]. Recently, Alhendal et al. reported a hybrid filtration cell (HFC), which utilizes limestone and activated carbons, for fluoride removal from water, with results demonstrating that fluoride could be completely removed from artificial water when the HFC is run at pH of 5.0, initial fluoride concentration of 30 mg/L and adsorbent dosage of 30 mg/L [104].

Phytoremediation is the direct application of green plants and their associated microorganisms to stabilize or absorb contaminants. A number of phytoremediation studies for removal of fluoride ions from water have posted impressive results with very high removal efficiencies. Aquatic plants, viz. *Pistia stratiotes*, *Eichhornia crassipes*, and *Spirodela polyrhiza* that grow in natural water bodies, have also been shown to remove fluoride ions from water with removal efficiencies of 19.87, 12.71 and 19.23%, respectively [105]. Fluoride-resistant bacteria play a major role in bioremediation and biotransformation of fluoride ions to convert them as less available and less toxic form and effectively reduce the F^- by binding them with ionophores [106]. Bioremediation potential by *Providencia vermicola* (KX926492) bacteria at 82% at pH 7, 37°C has been reported [107]. Elsewhere, encapsulated active growing blue-green algae, *Phormidium* sp., was used with a 60% removal efficiency from 3.0 mg/L initial fluoride concentration [98].

In a recent study by Maghanga et al. results reported indicated that *Maerua subcordata* root powder (MSRP) is a viable plant in defluoridation with ≈68% fluoride ion removal efficiency [101]. Additional related studies have reported the use of *Moringa oleifera* seeds which can be used for treatment of water containing fluorides as well as removal of turbidity through biocoagulation [100]. Other biocoagulants include neem leaves, tealeaves, tamarind seeds, banana peels, among many others [60, 96, 98, 99]. Suneetha et al. indicated effective adsorption of fluorides using active carbon derived from *Vitex negundo* plant [64]. Interesting studies on tamarind leaves and fruit pulp have shown its ability to facilitate the detoxification of fluoride from the body, in addition to the removal of fluoride ions from water (https://www.biologicalmedicineinstitute.com/post/tamarind-fluoride).

pH is a crucial component in adsorption studies, as high adsorption efficiencies are obtained at specific pH values, as shown in **Table 5**. Specifically, most adsorbents have reported a maximum fluoride removal at pH of 3–8. For example, whereas fluoride removal by tamarind is favored at neutral pH [96], at higher pH, hydrated cement defluoridation capacity remarkably reduces due to competition between F- and OH- [94]. Most of the adsorption remediation studies adopted Batch flow experiments to test the viability for real field water samples with optimization of other parameters such as initial concentration of fluoride, water temperature, contact time, rate of stirring and adsorbent dosage to reach the highest removal of fluorides.

7. Conclusion

Clean water and sanitation is one of the sustainable development goals (SDGs) that the world aspires to achieve by 2030. The challenge to this goal is the continued pollution of the environment by both natural and anthropogenic sources, thus limiting the available clean water. The technological advancement for addressing this issue has not been felt in all regions of the world, with the developing nations suffering the most. Pollution of water sources with fluoride is a major threat to the human well-being because it has been demonstrated that high fluoride levels result in the most severe forms of skeletal fluorosis that manifest in the form of disability. As a way of eliminating these pollutants, various technologies have been proposed, with green technologies being the most attractive because of the minimized negative impact to the environment. Such technologies, especially those that utilize naturally available materials, are attractive and can be easily adopted in a local setup for water purification toward addressing any form of fluorosis. As such, this calls for enhanced investment in the form of research to provide solutions to rural populations in developing nations where clean water supply from the local governments may be a challenge.

Acknowledgements

Veronica Okello acknowledges the Organization for Women in Science for the Developing World (OWSD) for their support through the Early Career Fellowship Programme grant (Award Agreement Number 4500429516). Elizabeth Ndunda is grateful to the Royal Society for their support through the FLAIR fellowship programme grant (Grant Number FLR\R1\192054).

Conflict of interest

The authors declare no conflict of interest.

Author details

Veronica A. Okello[1*], Elizabeth N. Ndunda[1], Abisaki Esitsakha[1] and Mercy Jeptoo[2]

1 Department of Physical Sciences, School of Pure and Applied Sciences, Machakos University, Machakos, Kenya

2 Department of Mining and Mineral Processing Engineering, Taita Taveta University, Voi, Kenya

*Address all correspondence to: vokello@mksu.ac.ke

References

[1] IAEA. All about water. IAEA Bulletin [Internet]. 2011;**53**(1):17-20. Available from: https://www.iaea.org/sites/default/files/publications/magazines/bulletin/bull53-1/53105911720.pdf. [Accessed: October 1, 2023]

[2] Annadurai ST, Rengasamy KJ, Sundaram R, Munusamy PA. Incidence and effects of fluoride in Indian natural ecosystem: A review. Advances in Applied Science Research. 2014;**5**(2):173-185

[3] Boretti A, Rosa L. Reassessing the projections of the world water development report. NPJ Clean Water. 2019;**2**:15. DOI: 10.1038/s41545-019-0039-9

[4] Unde MP, Patil RU, Dastoor PP. The untold story of fluoridation: Revisiting the changing perspectives. Indian Journal of Occupational and Environmental Medicine. 2018;**22**(3):121-127. DOI: 10.4103/ijoem.IJOEM_124_18

[5] CDC. Water Fluoridation Data & Statistics. Centre for Disease Control and prevention. 2020. Available from: https://www.cdc.gov/fluoridation/statistics/index.htm [Accessed: May 9, 2023]

[6] Coplan MJ, Patch SC, Masters RD, Bachman MS. Confirmation of and explanations for elevated blood lead and other disorders in children exposed to water disinfection and fluoridation chemicals. Neurotoxicology. 2007;**28**:1032-1042. DOI: 10.1016/j.neuro.2007.02.012

[7] Onipe T, Edokpayi JN, Odiyo JO. A review on the potential sources and health implications of fluoride in groundwater of sub-Saharan Africa. Journal of Environmental Science and Health - Part A. 2020;**55**(9):1078-1093. DOI: 10.1080/10934529.2020.1770516

[8] Mwiathi NF, Gao X, Li C, Rashid A. The occurrence of geogenic fluoride in shallow aquifers of Kenya Rift Valley and its implications in groundwater management. Ecotoxicology and Environmental Safety. 2022;**229**:113046. DOI: 10.1016/j.ecoenv.2021.113046

[9] Greenwood NN, Earnshaw A. The halogens: Fluorine, chlorine, bromine, iodine and astatine. In: Chemistry of the Elements. 2nd ed. UK: University of Leeds; 1997. DOI: 10.1016/b978-0-7506-3365-9.50023-7

[10] Singh G, Kumari B, Sinam G, Kriti KN, Mallick S. Fluoride distribution and contamination in the water, soil and plants continuum and its remedial technologies, an Indian perspective – A review. Environmental Pollution. 2018;**239**:95-108. DOI: 10.1016/j.envpol.2018.04.002

[11] Podgorski J, Berg M. Global analysis and prediction of fluoride in groundwater. Nature Communications. 2022;**13**(1):1-9. DOI: 10.1038/s41467-022-31940-x

[12] Kut KMK, Sarswat A, Srivastava A, Pittman CU, Mohan D. A review of fluoride in African groundwater and local remediation methods. Groundwater for Sustainable Development. 2026;**2-3**:190-212. DOI: 10.1016/j.gsd.2016.09.001

[13] Meenakshi MRC. Fluoride in drinking water and its removal. Journal of Hazardous Materials. 2006;**137**(1):456-463. DOI: 10.1016/j.jhazmat.2006.02.024

[14] Manji F, Baelum V, Fejerskov O. Dental fluorosis in an area of Kenya with 2 ppm fluoride in the drinking water. Journal of Dental Research. 1986;**65**(5):659-662. DOI: 10.1177/00220345860650050501

[15] Gevera PK, Cave M, Dowling K, Gikuma-Njuru P, Mouri H. A review on the occurrence of some potentially harmful elements in the natural environment and their health implications: Examples of fluoride, iron and salinity in the southern-eastern Kenya region. In: Siegel M, Selinus O, Finkelman R, editors. Practical Applications of Medical Geology. UK: University of Edinburgh; 2021. pp. 637-670. DOI: 10.1007/978-3-030-53893-4_19

[16] Gevera PK, Cave M, Dowling K, Gikuma-njuru P, Mouri H. Naturally occurring potentially harmful elements in groundwater in makueni county, South-Eastern Kenya: Effects on drinking water quality and agriculture. Geosciences. 2020;**10**(2):1-23. DOI: 10.3390/geosciences10020062

[17] Mbithi FM, Patrick CK, Peter GN. Assessment of the impact of groundwater fluoride on human health: A case study of Makindu District in Kenya. Journal of Earth Science & Climatic Change. 2017;**8**(4):396-402. DOI: 10.4172/2157-7617.1000396

[18] Gevera PK, Cave M, Dowling K, Gikuma-Njuru P, Mouri H. Potential fluoride exposure from selected food crops grown in high fluoride soils in the Makueni County, South-Eastern Kenya. Environmental Geochemistry and Health. 2022;**44**(12):4703-4717. DOI: 10.1007/s10653-022-01240-w

[19] Mutiso SK, Thompson RD. Future water supply in the Machakos District of Kenya. Ambio. 1987;**16**(6):322-325. Available from: https://www.jstor.org/stable/4313394

[20] Muraga JM. Assessment of Dissolved Ions and Microbial Coliforms in Water from Selected Sites of the Upper Athi River Subcatchment Area, Kenya. Kenya: University of Nairobi; 2010

[21] Nzeve JK, Mbate GN. Assessment of Ground water quality in Mumbuni estate, Machakos town, Kenya. Journal of Health and Environmental Research. 2021;**7**(1):7-11. DOI: 10.11648/j.jher.20210701.12

[22] James MR. Dental Fluorosis and Parental Knowledge of Risk Factors for Dental Fluorosis. University of the Western Cape; 2016. Available from: https://core.ac.uk/download/pdf/58916093.pdf

[23] Nzeve JK, Matata ACA. Assessment of water quality in Kalundu stream and Kalundu dam in Kitui County, Kenya. Journal of Environment and Earth Science. 2021;**11**(1):36-40. DOI: 10.7176/jees/11-1-05

[24] Wambua T, Hunja CW, Kimatu J, Muluvi GM, Kitheka JU, Mutiso F, et al. A comparative study of the physiochemical and bacteriological parameters of potable water from different sources in Kitui County, Kenya. Journal of Environment and Earth Science. 2022;**12**(11):48-56. DOI: 10.7176/jees/12-11-05

[25] Gevera P, Mouri H, Maronga G. Occurrence of fluorosis in a population living in a high-fluoride groundwater area: Nakuru area in the central Kenyan Rift Valley. Environmental Geochemistry and Health. 2019;**41**:829-840. DOI: 10.1007/s10653-018-0180-2

[26] Rajkovic M, Novakovic I. Determination of fluoride content in drinking water and tea infusions using fluoride ion selective electrode. Journal of Agricultural Sciences, Belgrade. 2007;**52**(2):155-168. DOI: 10.2298/jas0702155r

[27] Okamoto Y, Yasukawa N, Fujiwara T, Iwamoto E. Determination of fluoride ion in aqueous samples by inductively coupled plasma atomic emission spectrometry with tungsten boat furnace vaporizer. Journal of Analytical Atomic

Spectrometry. 2000;**15**(7):809-811. DOI: 10.1039/b003036h

[28] Schaumburg K, Deverel C. Fluorine-19 nuclear magnetic resonance chemical shift. Journal of the American Chemical Society. 1967;**1879**(9):2495-2499. DOI: 10.1021/ja01012a009

[29] Barghouthi Z, Amereih S. Spectrophotometric determination of fluoride in groundwater using resorcin blue complexes. American Journal of Analytical Chemistry. 2012;**3**(09):651-655. DOI: 10.4236/ajac.2012.39085

[30] Du J, Sheng C, Wang Y, Zhang H, Jiang K. Determination of trace fluoride in water samples by silylation and gas chromatography/mass spectrometry analysis. Rapid Communications in Mass Spectrometry. 2021;**35**(11):1-6. DOI: 10.1002/rcm.9089

[31] Alhaddad M, El-sheikh SM. Selective and fast detection of fluoride-contaminated water based on a novel Salen-Co-MOF Chemosensor. ACS Omega. 2021;**6**(23):15182-15191. DOI: 10.1021/acsomega.1c01424

[32] Sharafizadeh M, Mokhtari J. Anion recognition by urea metal – Organic frameworks: Remarkable sensitivity for arsenate and fluoride ions. Environmental Science and Pollution Research. 2020;**27**:25132-25139. DOI: 10.1007/s11356-020-08934-7

[33] Xie D, Ge X, Qin W, Zhang Y. NH_2-MIL-53 (Al) for simultaneous removal and detection of fluoride. Chinese Journal of Chemical Physics. 2021;**34**(2):227-237. DOI: 10.1063/1674-0068/cjcp2004054

[34] Sun Y, Xu X, Zhao Y, Tan H, Li Y, Du J. Luminescent metal organic frameworks–based chemiluminescence resonance energy transfer platform for turn–on detection of fluoride ion. Talanta. 2020;**209**:120582. DOI: 10.1016/j.talanta.2019.120582

[35] Mantasha I, Shahid M, Saleh HAM, Qasem KMA, Ahmad M. A novel sustainable metal organic framework as ultimate aqueous phase sensor for natural hazards: Detection of nitrobenzene and F^- at ppb level and rapid and selective adsorption of methylene blue. CrystEngComm. 2020;**22**(22):3891-3909. DOI: 10.1039/D0CE00356E

[36] Zhu H, Huang J, Zhou Q, Lv Z, Li C, Hu G. Enhanced luminescence of NH_2 -UiO-66 for selectively sensing fluoride anion in water medium. Journal of Luminescence. 2019;**208**:67-74. DOI: 10.1016/j.jlumin.2018.12.007

[37] Wang X, Chu C, Wu Y, Deng Y, Zhou J, Yang M, et al. Synthesis of yttrium (III)-based rare-earth metal-organic framework nanoplates and its applications for sensing of fluoride ions and pH. Sensors & Actuators: B Chem. 2020;**321**:128455. DOI: 10.1016/j.snb.2020.128455

[38] Thavarajah W, Silverman AD, Verosloff MS, Kelley-Loughnane N, Jewett MC, Lucks JB. Point-of-use detection of environmental fluoride via a cell-free riboswitch-based biosensor. ACS Synthetic Biology. 2020;**9**(1):10-18. DOI: 10.1021/acssynbio.9b00347

[39] Mukherjee S, Shah M, Chaudhari K, Jana A, Sudhakar C, Srikrishnarka P, et al. Smartphone-based fluoride-specific sensor for rapid and affordable colorimetric detection and precise quantification at sub-ppm levels for field applications. ACS Omega. 2020;**5**(39):25253-25263. DOI: 10.1021/acsomega.0c03465

[40] Jiménez-Becerril J, Solache-Ríos M, García-Sosa I. Fluoride removal from aqueous solutions by boehmite. Water, Air, and Soil Pollution. 2012;**223**(3):1073-1078. DOI: 10.1007/s11270-011-0925-3

[41] Fordyce FM. Fluorine: Human health risks. In: Nriagu JO, editor. Encyclopedia of Environmental Health. Vol. 2. Edinburgh, UK: British Geological Survey; 2011. pp. 776-785

[42] Kanduti D, Sterbenk P, Artnik B. Fluoride: A review of use and effects on health. Materia Socio Medica. 2016;**28**(2):133-137. DOI: 10.5455/msm.2016.28.133-137

[43] Levy M, Corbeil F. In: Levy M, Corbeil F, Fortin C, Lamarre J-R, Lavalliere A, Schwartz S, Tardif R, Vu D-D, editors. Water Fluoridation: An Analysis of the Health Benefits and Risks. Quebec Public Health Institute; 2007. Available from: https://centerforinquiry.org/wp-content/uploads/sites/33/quackwatch/quebec_2007.pdf

[44] Li Y, Liang C, Slemenda CW, Ji R, Sun S, Cao J, et al. Effect of long-term exposure to fluoride in drinking water on risks of bone fractures. Journal of Bone and Mineral Research. 2001;**16**(5):932-939. DOI: 10.1359/jbmr.2001.16.5.932

[45] Jha SK, Singh RK, Damodaran T, Mishra VK, Sharma DK, Rai D. Fluoride in groundwater: Toxicological exposure and remedies. Journal of Toxicology and Environmental Health - Part B: Critical Reviews. 2013;**16**(1):52-66. DOI: 10.1080/10937404.2013.769420

[46] Yadav KK, Kumar S, Pham QB, Gupta N, Rezania S, Kamyab H, et al. Fluoride contamination, health problems and remediation methods in Asian groundwater: A comprehensive review. Ecotoxicology and Environmental Safety. 2019;**182**(1):109362. DOI: 10.1016/j.ecoenv.2019.06.045

[47] Yadav M, Singh G, Jadeja RN. Fluoride contamination in groundwater, impacts, and their potential remediation techniques. In: Madhav S, Singh P, editors. Groundwater Geochemistry: Pollution and Remediation Methods. Chichester, Great Britain: John Wiley & Sons Ltd; 2021. pp. 22-41. DOI: 10.1002/9781119709732.ch2

[48] Davies BE, Bowman C, Davies TC, Selinus O. Medical geology: Perspectives and prospects. In: Selinus O, editor. Essentials of Medical Geology. Dordrecht: Springer; 2013. DOI: 10.1007/978-94-007-4375-5_1

[49] Thole B. Ground water contamination with fluoride and potential fluoride removal technologies for east and southern Africa. In: Ahmad I, Dar MA, editors. Perspectives in Water Pollution. London, UK: IntechOpen; 2013. pp. 65-99. DOI: 10.5772/54985

[50] Shahab S, Mustafa G, Khan IA, Zahid M. Effects of fluoride ion toxicity on animals, plants, and soil health: A review. Fluoride. 2017;**50**(4):393-408

[51] Hong B, Joo R, Lee K, Lee D, Rhie J, Min S. Fluoride in soil and plant. Korean Journal of Agricultural Science. 2016;**43**(4):522-536. DOI: 10.7744/kjoas.20160054

[52] Everett ET. Fluoride's effects on the formation of teeth and bones, and the influence of genetics. Journal of Dental Research. 2011;**90**(5):552-560. DOI: 10.1177/0022034510384626

[53] Menya D, Maina SK, Kibosia C, Kigen N, Oduor M, Some F, et al. Dental fluorosis and oral health in the African Esophageal cancer corridor: Findings from the Kenya ESCCAPE case – Control study and a pan-African perspective. International Journal of Cancer. 2019;**145**:99-109. DOI: 10.1002/ijc.32086

[54] Demarchi P, Garbarino F, Mascolo A, Biavati FS, Ugolini A. Fluorosis and Oral health status in adolescents living in a high-fluoride groundwater area: A case study of Nairobi suburbs (Kenya).

Applied Sciences. 2022;**13**(1):368. DOI: 10.3390/app13010368

[55] Wambu EW, Agong SG, Anyango B, Akuno W, Akenga T. High fluoride water in Bondo-Rarieda area of Siaya County, Kenya: A hydro-geological implication on public health in the Lake Victoria Basin. BMC Public Health. 2014;**14**:462-470. DOI: 10.1186/1471-2458-14-462

[56] Ayoob S, Gupta AK, Bhat VT. A conceptual overview on sustainable technologies for the defluoridation of drinking water. Critical Reviews in Environmental Science and Technology. 2008;**38**(6):401-470. DOI: 10.1080/10643380701413310

[57] Gupta VK, Ali I. Environmental Water: Advances in Treatment, Remediation and Recycling. 1st ed. Amsterdam Netherlands: Elsevier; 2013

[58] Onyango MS, Kojima Y, Aoyi O, Bernardo EC, Matsuda H. Adsorption equilibrium modeling and solution chemistry dependence of fluoride removal from water by trivalent-cation-exchanged zeolite F-9. Journal of Colloid and Interface Science. 2004;**279**(2):341-350. DOI: 10.1016/j.jcis.2004.06.038

[59] Zhao M, Wang Q, Krua LSN, Yi R, Zou R, Li X, et al. Application Progress of new adsorption materials for removing fluorine from water. Water. 2023;**15**(4):646. DOI: 10.3390/w15040646

[60] Bharali RK, Bhattacharyya KG. Biosorption of fluoride on neem (Azadirachta indica) leaf powder. Journal of Environmental Chemical Engineering. 2015;**3**(2):662-669. DOI: 10.1016/j.jece.2015.02.007

[61] Yadav AK, Abbassi R, Gupta A, Dadashzadeh M. Removal of fluoride from aqueous solution and groundwater by wheat straw, sawdust and activated bagasse carbon of sugarcane. Ecological Engineering. 2013;**52**:211-218. DOI: 10.1016/j.ecoleng.2012.12.069

[62] Jiang C, Cui S, Han Q, Li P, Zhang Q, Song J, et al. Study on application of activated carbon in water treatment. IOP Conference Series. 2019;**237**:022049. DOI: 10.1088/1755-1315/237/2/022049

[63] Bakhta S, Sadaoui Z, Bouazizi N, Samir B, Allalou O, Devouge-Boyer C, et al. Functional activated carbon: From synthesis to groundwater fluoride removal. RSC Advances. 2022;**12**(4):2332-2348. DOI: 10.1039/d1ra08209d

[64] Suneetha M, Sundar BS, Ravindhranath K. Removal of fluoride from polluted waters using active carbon derived from barks of Vitex negundo plant. Journal of Analytical Science and Technology. 2015;**6**(15):1-19. DOI: 10.1186/s40543-014-0042-1

[65] Liu M, Xiao C. Research progress on modification of activated carbon. E3S Web of Conferences. 2018;**38**:02005. DOI: 10.1051/e3sconf/20183802005

[66] Fito J, Said H, Feleke S, Worku A. Fluoride removal from aqueous solution onto activated carbon of Catha edulis through the adsorption treatment technology. Environmental Systems Research. 2019;**8**(25):1-10. DOI: 10.1186/s40068-019-0153-1

[67] Reynel-AvilaHE,Camacho-AguilarKI, Bonilla-PetricioletA,Mendoza-CastilloDI, Gonzalez-Ponce HA, Trejo-Valencia R. Review article engineered magnetic carbon-based adsorbents for the removal of water priority pollutants: An overview. Adsorption Science & Technology. 2021;**2021**:1-41. DOI: 10.1155/2021/9917444

[68] Waghmare SS, Arfin T. Fluoride removal from water by calcium materials: A state-of-the-art review. International Journal of Innovative Research in Science, Engineering and Technology. 2015;**4**(9):8090-8102. DOI: 10.15680/ijirset.2015.0409013

[69] Chen W, Wu Y, Xie Z, Li Y, Tang W, Yu J. Calcium hydroxide recycled from waste eggshell resources for the effective recovery of fluoride from wastewater. RSC Advances. 2022;**12**(43):28264-28278. DOI: 10.1039/d2ra05209a

[70] Wallace AR, Su C, Choi Y-K, Kan E, Sun W. Removal of fluoride from water using a calcium-modified dairy manure–derived biochar. Journal of Environmental Engineering. 2020;**146**(12):1-26. DOI: 10.1061/(asce)ee.1943-7870.0001812

[71] Alkurdi SSA, Al-Juboori RA, Bundschuh J, Hamawand I. Bone char as a green sorbent for removing health threatening fluoride from drinking water. Environment International. 2019;**127**:704-719. DOI: 10.1016/j.envint.2019.03.065

[72] Singh S, German M, Chaudhari S, Sengupta AK. Fluoride removal from groundwater using zirconium impregnated anion exchange resin. Journal of Environmental Management. 2020;**263**(2):110415. DOI: 10.1016/j.jenvman.2020.110415

[73] Goswami A, Purkait MK. The defluoridation of water by acidic alumina. Chemical Engineering Research and Design. 2012;**90**(12):2316-2324. DOI: 10.1016/j.cherd.2012.05.002

[74] Hu X, Peng X, Kong L. Removal of fluoride from zinc sulfate solution by in situ Fe(III) in a cleaner desulfuration process. Journal of Cleaner Production. 2017;**164**:163-170. DOI: 10.1016/j.jclepro.2017.06.213

[75] Tripathy SS, Raichur AM. Abatement of fluoride from water using manganese dioxide-coated activated alumina. Journal of Hazardous Materials. 2008;**153**(3):1043-1051. DOI: 10.1016/j.jhazmat.2007.09.100

[76] Dayananda D, Sarva VR, Prasad SV, Arunachalam J, Ghosh NN. Preparation of CaO loaded mesoporous Al_2O_3: Efficient adsorbent for fluoride removal from water. Chemical Engineering Journal. 2014;**248**:430-439. DOI: 10.1016/j.cej.2014.03.064

[77] Nazari M, Halladj R. Adsorptive removal of fluoride ions from aqueous solution by using sonochemically synthesized nanomagnesia/alumina adsorbents: An experimental and modeling study. Journal of the Taiwan Institute of Chemical Engineers. 2014;**45**(5):2518-2525. DOI: 10.1016/j.jtice.2014.05.020

[78] Maliyekkal SM, Sharma AK, Philip L. Manganese-oxide-coated alumina: A promising sorbent for defluoridation of water. Water Research. 2006;**40**(19):3497-3506. DOI: 10.1016/j.watres.2006.08.007

[79] Shi Q, Huang Y, Jing C. Synthesis, characterization and application of lanthanum-impregnated activated alumina for F removal. Journal of Materials Chemistry A. 2013;1(41):12797-12803. https://doi.org/10.1039/c3ta12548c

[80] Jin H, Ji Z, Yuan J, Li J, Liu M, Xu C, et al. Research on removal of fluoride in aqueous solution by alumina-modified expanded graphite composite. Journal of Alloys and Compounds. 2015;**620**:361-367. DOI: 10.1016/j.jallcom.2014.09.143

[81] Soni R, Modi S. Removal of fluoride from drinking water using red mud. International Journal of Scientific & Technology Research. 2013;**2**:120-122

[82] Vardhan VCM, Srimurali M. Removal of fluoride from water using a novel sorbent lanthanum-impregnated bauxite. Springerplus. 2016;**5**(1):1426-1444. DOI: 10.1186/s40064-016-3112-6

[83] Vinati A, Mahanty B, Behera SK. Clay and clay minerals for fluoride removal from water: A state-of-the-art review. Applied Clay Science. 2015;**114**:340-348. DOI: 10.1016/j.clay.2015.06.013

[84] Ingle N, Dubey H, Kaur N, Sharma I. Defluoridation techniques: Which one to choose. Journal of Health Research and Reviews. 2014;**1**(1):1-3. DOI: 10.4103/2394-2010.143315

[85] Jamwal KD, Slathia D. A review of Defluoridation techniques of global and Indian prominence. Current World Environment. 2022;**17**(1):41-57. DOI: 10.12944/cwe.17.1.5

[86] Figoli A, Hoinkis J, Bundschuh J. Membrane technologies for water treatment: Removal of toxic trace elements with emphasis on arsenic, fluoride and uranium. Johnson Matthey Technology Review. 2016;**60**(4):273-276. DOI: 10.1201/b19227

[87] Samrat MVN, Rao KK, SenGupta AK, Riotte J, Mudakavi JR. Defluoridation of reject water from a reverse osmosis unit and synthetic water using adsorption. Journal of Water Process Engineering. 2018;**23**:327-337. DOI: 10.1016/j.jwpe.2017.07.015

[88] Bey S, Criscuoli A, Simone S, Figoli A, Benamor M, Drioli E. Hydrophilic PEEK-WC hollow fibre membrane contactors for chromium (VI) removal. Desalination. 2011;**283**:16-24. DOI: 10.1016/j.desal.2011.04.038

[89] Feenstra L, Vasak L, Griffion J. Fluoride in Groundwater: Overview and Evaluation of Removal Methods. Delft, Netherlands: International Groundwater Resources Assessment Centre; 2007. Available from: https://www.un-igrac.org/sites/default/files/resources/files/IGRAC-SP2007-1_Fluoride-removal.pdf

[90] Katiyar P, Pandey N, Sahu KK. Biological approaches of fluoride remediation: Potential for environmental clean-up. Environmental Science and Pollution Research. 2020;**27**(12):13044-13055. DOI: 10.1007/s11356-020-08224-2

[91] Yadav AK, Kaushik CP, Haritash AK, Kansal A, Rani N. Defluoridation of groundwater using brick powder as an adsorbent. Journal of Hazardous Materials. 2006;**128**(2-3):289-293. DOI: 10.1016/j.jhazmat.2005.08.006

[92] Singh R, Raghuvanshi SP, Kaushik CP. Defluoridation of drinking water using brick powder as an adsorbent. Asian Journal of Chemistry. 2008;**20**(8):5818-5826

[93] Tarali SV, Hoolikantimath NP, Kulkarni N, Ghorpade PA. A novel cement-based technology for the treatment of fluoride ions. SN Applied Sciences. 2020;**2**(7):1-9. DOI: 10.1007/s42452-020-2986-7

[94] Kagne S, Jagtap S, Dhawade P, Kamble SP, Devotta S, Rayalu SS. Hydrated cement: A promising adsorbent for the removal of fluoride from aqueous solution. Journal of Hazardous Materials. 2008;**154**:88-95. DOI: 10.1016/j.jhazmat.2007.09.111

[95] Meenakshi S, Sundaram CS, Sukumar R. Enhanced fluoride sorption by mechanochemically activated kaolinites. Journal of Hazardous Materials. 2008;**153**(1-2, 172):164. DOI: 10.1016/j.jhazmat.2007.08.031

[96] Murugan M, Subramanian E. Studies on defluoridation of water by tamarind seed, an unconventional biosorbent. Journal of Water and Health.

2006;**4**(4):453-461. DOI: 10.2166/wh.2006.014

[97] Mittal Y, Srivastava P, Kumar N, Yadav KA. Remediation of fluoride contaminated water using encapsulated active growing blue-green algae. Phormidium sp. Environmental Technology & Innovation. 2020;**19**:100855. DOI: 10.1016/j.eti.2020.100855

[98] Mondal NK, Roy A. Potentiality of a fruit peel (banana peel) toward abatement of fluoride from synthetic and underground water samples collected from fluoride affected villages of Birbhum district. Applied Water Science. 2018;**8**(3):1-10. DOI: 10.1007/s13201-018-0729-3

[99] Cai H, Chen G, Peng C, Zhang Z, Dong Y, Shang G, et al. Removal of fluoride from drinking water using tea waste loaded with Al/Fe oxides: A novel, safe and efficient biosorbent. Applied Surface Science. 2014;**328**:34-44. DOI: 10.1016/j.apsusc.2014.11.164

[100] Ravikumar K, Sheeja AK. Removal of fluoride from aqueous system using Moringa oleifera seed cake and its composite coagulants. International Journal of Current Engineering Technology. 2014;**4**(3):1356-1360

[101] Maghanga J, Okello V, Michira J, Ojwang L, Mati B, Segor F. Fluoride in water, health implications and plant-based remediation strategies. Physical Sciences Reviews. 2022:1-19. DOI: 10.1515/psr-2022-0123

[102] Obijole O, Mugera GW, Mudzielwana R, Ndungu P, Samie A, Babatunde A. Hydrothermally treated aluminosilicate clay (HTAC) for remediation of fluoride and pathogens from water: Adsorbent characterization and adsorption modelling. Water Resources and Industry. 2021;**25**:100144. DOI: 10.1016/j.wri.2021.100144

[103] Cherukumilli KAK, Delaire C, Amrose SE, Gadgil AJ. Factors governing the performance of bauxite for fluoride remediation of groundwater factors governing the performance of bauxite for fluoride remediation of groundwater. Environmental Science and Technology. 2017;**51**(4):2321-2328. DOI: 10.1021/acs.est.6b04601

[104] Alhendal M, Nasir MJ, Hashim KS, Amoako-Attah J, Muradov M, Kot P, et al. Cost-effective hybrid filter for remediation of water from fluoride cost-effective hybrid filter for remediation of water from fluoride. IOP Conference Series: Materials Science and Engineering. 2020;**888**:012038. DOI: 10.1088/1757-899X/888/1/012038

[105] Karmakar S, Mukherjee J, Mukherjee S. Removal of fluoride contamination in water by three aquatic plants. International Journal of Phytoremediation. 2015;**18**(3):222-227. DOI: 10.1080/15226514.2015.1073676

[106] Chatterjee N, Sahu G, Bag AG, Plal B, Hazra GC. Role of fluoride on soil, plant and human health: A review on its sources, toxicity and mitigation strategies. International Journal of Environment and Climate Change. 2020;**10**(8):77-90. DOI: 10.9734/ijecc/2020/v10i830220

[107] Mukherjee S, Sahu P, Halder G. Microbial remediation of fluoride-contaminated water via a novel bacterium Providencia vermicola (KX926492). Journal of Environmental Management. 2017;**204**:413-423. DOI: 10.1016/j.jenvman.2017.08.051

Chapter 2

Reuse of Treated Water from Municipal Treatment Plants in Mexico

Ángeles Martínez-Orgániz, Ricardo Herrera-Navarrete, Daniel Pineda-Mora, Viridiana Del Carmen-Niño and Rosa Iris Balbuena-Hernández

Abstract

Wastewater treatment plants (WWTPs) receive a wide variety of contaminants that cannot be eliminated or completely removed with current conventional methods. In this sense, the development and use of advanced technologies is a challenge in countries where wastewater sanitation is hardly a guarantee. However, the reuse of treated urban wastewater can function as an alternative to mitigate water pressure and, at the same time, guarantees water quality for potential reuse in agriculture, in the irrigation of landscape or urban green areas, but especially for aquifer recharge. Therefore, this chapter is focused on reviewing the current state of WWTPs in Mexico and the potential reuse of treated water.

Keywords: water quality, water reuse, aquifer recharge, emerging contaminants, Acapulco

1. Introduction

Water is an indispensable natural resource for promoting economic and social development, as well as being of vital and critical importance for ecosystems [1, 2]. Rapid population growth, high rates of urbanization, and climate change are key factors that put pressure on water resources [3]. Of the approximately 1388 million km^3 of water on the planet, only 3% (41.64 million km^3) is freshwater and less than 1% (13.88 million km^3) is accessible for human consumption [4], with availability limited by quality. According to Wang et al. [5], millions of people die each year from water pollution-related diseases, and they estimate that by 2050 more than half of the world's population will live in water-scarce regions [6].

In addition to water scarcity, pollution has become a matter of global interest and concern [6], and even more so when considering the appearance of new pollutants called emerging contaminants (ECs), identified mainly in urban wastewater, a consequence of the consumption habits of modern society [7]. In this context, the role of wastewater treatment plants (WWTPs) to face the growing need for larger volumes of contaminant-free water, whose main objective is to ensure the safety of human health and environmental protection, trying to achieve sustainable urban development of local waters, stands out [8].

Conventional WWTPs consist of four stages for efficient treatment (preliminary, primary, secondary, and tertiary) [9]. However, most of these units fail to remove a wide range of emerging contaminants present in wastewater [10]. Currently, the development of new technologies can be considered as a viable alternative for the efficient processing of treated wastewater and potential reuse in water-demanding activities such as landscape irrigation, industry, agriculture, and even aquifer recharge [11].

Latin America is the region with the highest availability of freshwater; it has 33% of renewable water resources, although only part of it is accessible to the population [12]. In this part of the world, only 8% of the wastewater produced daily is treated. Another part is discharged into surface waters, and yet another part is used for irrigation, covering about 500,000 hectares, mostly with untreated water [13]. Countries such as China, Mexico, and the United States have been identified as those that reuse a large volume of wastewater for agricultural activities, and in some cases no efficient treatment is carried out [14]. In contrast, in other cities such as Windhoek, Namibia, and Orange Country, California, good practices in the reuse of treated wastewater (drinking water supply) have been documented [15].

According to Ghafoori et al. [16], Mexico uses wastewater to irrigate approximately 260,000 hectares of green areas (gardens), which is why this country continues to promote resource management for the construction, rehabilitation, maintenance, and operation of WWTPs [17]. On the other hand, Mexico has a comprehensive and modern regulatory framework that offers the possibility of recharging aquifers with reclaimed water under regulated guidelines, since it has a production potential of 144.7 m^3/s through its 2786 installed WWTPs, among which activated sludge treatment predominates as the most widely used process (69.6%) [15, 18].

Some WWTPs in Mexico operate in optimal conditions according to different studies carried out by the scientific community and governmental institutions in the country [18–20]. Treated wastewater is an underutilized resource despite being considered as a viable alternative in a context of environmental degradation in various bodies of water. Therefore, this study was focused on reviewing the current state of WWTPs in Mexico and the potential reuse of treated water in one of its main municipalities, Acapulco. This port located in the South of Mexico has 18 WWTPs, and its effluents are discharged without any use to the Pacific coast and the city's main river. This chapter aims to strengthen Mexico's commitment to the 2030 Agenda and to cover at the local-level Goal 6 (ensure availability and sustainable management of water and sanitation for all), which includes in its third target "halve the proportion of untreated wastewater and substantially increase recycling and safe reuse." Finally, this study may allow other research to expand on the issues related to reuse of treated wastewater and may serve as a guide for making better decisions on the applications of treated wastewater in areas with greater water stress in Mexico.

2. Perspective of wastewater management in Mexico

2.1 Water pollution

The problem of water pollution began to be noticed in the early nineteenth century and has become one of the most serious environmental problems of our time, generating a global scarcity of clean water, hence the importance of conserving and maintaining the quality of natural water sources to ensure its sustainability and use for future generations [21]. It is of utmost importance to consider adequate wastewater treatment processes that comply with the required parameters for the different types of reuse [15]. Mexico's water resources are facing serious pollution problems due to the fact that water quality is below the permissible limits for human health, both surface water and groundwater are used as receiving bodies for heavy loads of conventional and nonconventional pollutants [22].

In recent decades, impacts to the aquatic environment have been detected by nonconventional pollutants, called emerging contaminants (ECs), and these compounds of different origin and chemical nature are originated by drugs, pesticides, surfactants, surfactants, surfactants, personal care products, and among others and have raised great concern to the scientific community, due to the environmental damage they generate by their physicochemical characteristics when combined with other substances, including water and bioacomulative through the trophic chain [23]. Currently, ECs are not considered in monitoring or regulatory programs, despite the existence of several studies on their occurrence, fate, behavior, and toxicological effects on terrestrial and aquatic ecosystems [24]. The overuse and misuse of antibiotics have become a global problem, as the discharge of antibiotics can not only chemically contaminate water but also induce antibiotic-resistant bacteria (ARB) and antibiotic-resistant genes (ARGs) [10].

Currently, more than 600 active pharmaceutical substances (metabolites and transformation products) have been detected in the aquatic environment belonging to different therapeutic groups worldwide [25, 26]. The widespread use and abuse of active substances have made it possible to be detected in different environmental matrices (e.g. surface water, groundwater, wastewater, and stormwater runoff in urban areas) in various concentrations [27–29]. Therefore, understanding the fate of these pollutants in wastewater treatment can contribute to better management and improve the quality of treated effluents [30]. In Mexico, several studies have been conducted on the presence of ECs in aquatic ecosystems [31–36]. Water quality problems are severe and have a significant lag in their attention compared to those related to quantity and provision of services to the population. Water quality monitoring is a process that must be effective, regulated, and updated. In the same way, water quality assessment is essential to guide efforts to promote water reuse [37].

2.2 Characterization of wastewater

Rapid population growth and high urbanization rates represent challenges in water management; among them, the increase in wastewater generation [38]. Wastewater (WW) consists of 99% water and 1% suspended, dissolved, and colloidal solids [4]. According to the National Water Law of Mexico, WW is "a varied composition resulting from discharges of urban, domestic, industrial, commercial, service, agricultural, livestock, treatment plants and, in general, from any use, as well as a mixture of them" [39]. Other important sources of pollution to consider are hospitals and clinics, where hazardous waste is disposed into municipal sewers [38].

According to Valdes et al. [2], 135,600 liters per second (L/s) of wastewater were treated in Mexico, which corresponds to 63% of the total water recovered from the country's sewerage systems. An estimated reuse rate of 39,800 L/s was also documented directly from treatment plants and 78,800 L/s indirectly after its first discharge into a water body. However, wastewater discharges cause a high

Conventional contaminants	Treatment
Suspended solids	Sedimentation, roughing, filtration, and flotation. Addition of polymers or chemical reagents, coagulation-sedimentation.
Biodegradable organic matter	Activated sludge, fixed film: trickling filters, fixed film: biological disks, lagooning variations, intermittent sand filtration, and physicochemical systems.
Pathogens	Chlorination, hypochlorination, ozonation, and UV radiation.
Nitrogen	Variations of suspended cultivation systems (nitrification denitrification), variations of fixed film systems (nitrification denitrification), and ammonia entrainment (ion exchange).
Phosphorus	Addition of metallic salts, coagulation and sedimentation with salt, and biological removal of phosphorus.
Refractory organic matter	Carbon adsorption and tertiary ozonation.
Heavy metals	Chemical precipitation. Ion exchange.
Dissolved inorganic solids	Ion exchange, reverse osmosis, and electrodialysis.

Table 1.
Main conventional contaminants present in wastewater and their treatment [41].

Network	Area	Number of monitoring sites
Surface	Surface	2512
Underground	Underground	1060
Special studies	Underground bodies of water	49
	Surface bodies of water	156
Discharges	Underground	8
	Surface	429
Coastal	Coastal	820
	Total	5034

Table 2.
Sites of the water quality monitoring network in Mexico [43].

degree of stress on aquatic ecosystems [40], due to the presence of various types of contaminants resulting from inadequate treatment. In this sense, **Table 1** shows the behavior of conventional-type contaminants, which is unpredictable during treatment since it can be very effective for some of them and null for others [41].

2.3 Water quality assessment

Water quality is assessed from physical, chemical, and biological characteristics, evaluated individually or as a group. Physicochemical parameters give extensive information on the nature of the chemical species in the water and their physical properties; these analyses are rapid and can be monitored frequently [42]. In Mexico, the water authority (CONAGUA) has a National Water Quality Measurement Network (**Table 2**).

The main objectives of the network are (I) to provide the water authority and users with reliable results, which can be transformed into information for decision-making, and (II) to obtain water quality results from more than 5000 monitoring sites with the highest quality standards. The criteria for the selection of sites to assess water quality are based mainly on representativeness, standardization, and reliability, considering sources of contamination, reference sites and/or areas (hydrological basins), and among others.

To assess surface water quality, the following indicators are taken as the main reference: Biochemical Oxygen Demand (DBO_5), Chemical Oxygen Demand (COD), Total Suspended Solids (TSS), *Fecal coliforms* (FC), *Fecal enterococci* (FE), *Escherichia coli* (*E. coli*), and Dissolved Oxygen Saturation (DO), taking the median of the set of data from each study site. Whereas, for acute toxicity, it is calculated as the maximum of the toxicities with Daphnia magna and Vibrio fischeri. As shown in **Figure 1**, water quality results are marked in red or yellow when one or more indicators are not met and marked in green when all water quality indicators are met.

The results presented in **Figure 1** show that of the 4233 study sites only 30% (1266 sites marked in red) did not meet the following parameters: DBO_5, COD, toxicity, and/or *fecal Enterococci*. While 29.1% (1228 sites marked in yellow) do not comply with the following parameters: *E. coli*, *fecal coliforms*, TSS, and/or percentage of DO. In addition, only 40.9% (1727 sites marked in green) met all water quality indicators in Mexico [43]. Although the water quality indicators assessed in Mexico can provide a lot of information about the state of water resources, they exhibit at least two important problems: water scarcity and contamination; however, the latter does not

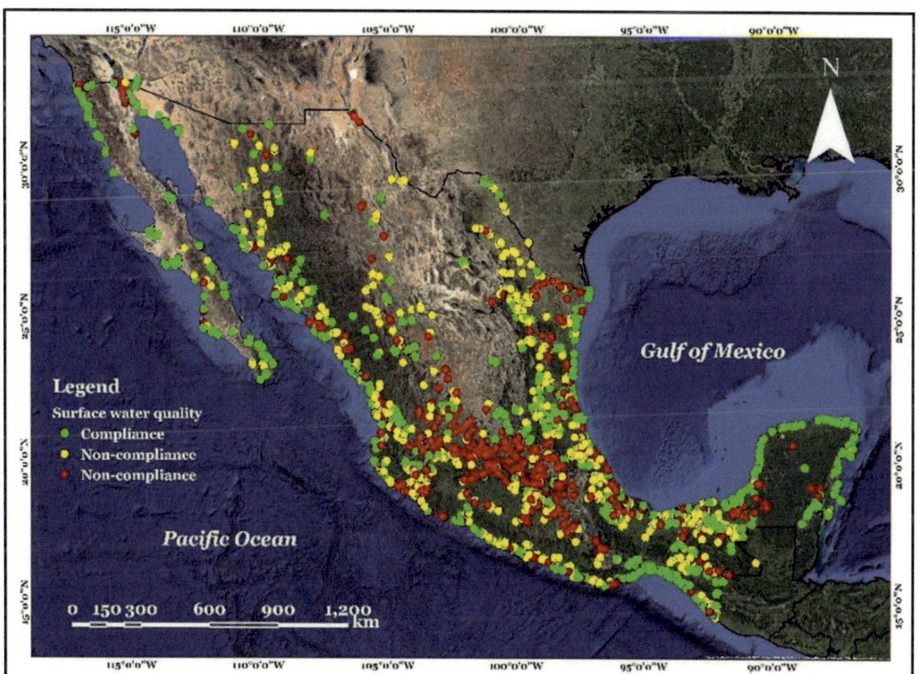

Figure 1.
Study sites for the assessment of surface water quality in Mexico [43].

consider the existence of multiple contaminants that, even in small quantities, can be harmful to health and/or the environment [43].

2.4 Sanitation systems in Mexico

2.4.1 Wastewater treatment plants (WWTPs)

The technologies commonly employed in WWTPs started globally in the early twentieth century; however, today they work with low levels of efficiency and high levels of energy consumption [44]. This situation requires considerable financial investments to renew and update the infrastructures with a focus on new socioeconomic paradigms such as the circular economy that require better use and reuse of water resources [3].

There are two types of wastewater treatment systems: centralized and decentralized. The centralized system is more common in developed countries where economies of scale favor large facilities, while the decentralized system is more attractive in developing countries, which produces lower energy use and simpler designs, but still represent high operating costs for local governments. For example, Mexico has a large urban and rural area that depends on decentralized systems for wastewater treatment [45].

There are 2786 WWTPs in the country (**Figure 2**), which treat approximately 65.7% of the wastewater produced, 144.71 m^3/s. Wastewater treatment does not exceed 70% of the water collected in the drainage systems. Sewerage coverage represents 97.39% in urban areas and 77.52% in rural areas. Of which, 58.8% (1637 WWTPs) have a treatment flow of <5 l/s, 13% (363 WWTPs) have a flow between

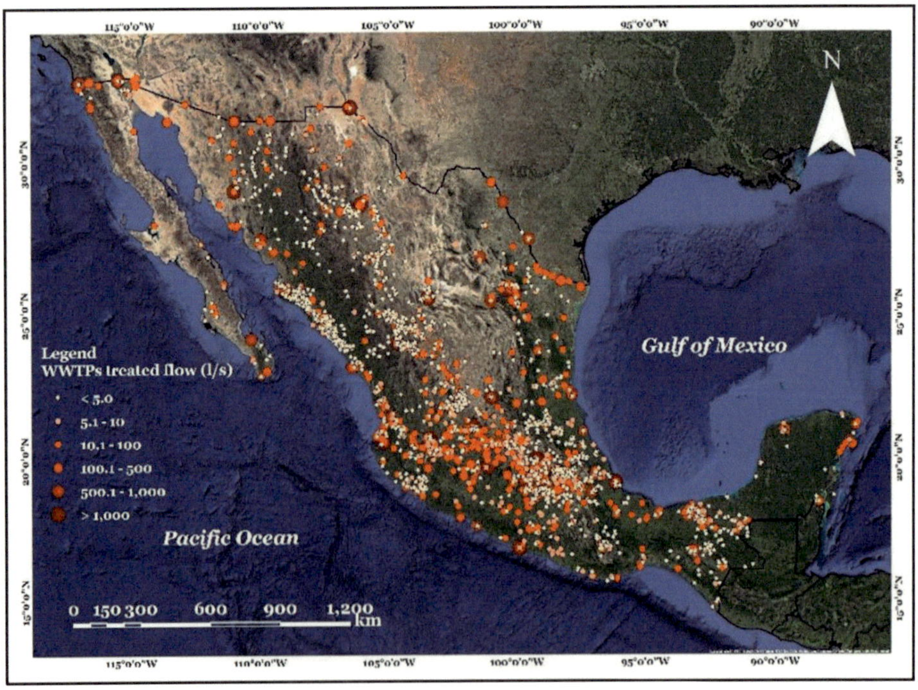

Figure 2.
Wastewater treatment plants in Mexico classified by treated flow (l/s) [43].

5.1 and 10 l/s, 22% (612 WWTPs) have a treatment flow of 10.1–100 l/s, 4.5% (125 WWTPs) have a treatment flow rate of 100.1–500 l/s, 0.8% (23 WWTPs) have a treatment flow rate of 500.1–1000 l/s, and 0.9% (26 WWTPs) have a treatment flow >1000 l/s [43].

The most recent information indicates that, until 2020, the country generated an approximate total volume of municipal wastewater of 8.82 thousand hm^3/year (279.80 m^3/s). Of this generated volume, only 6.79 thousand hm^3/year (215.40 m^3/s) were collected by the sewage systems. This means that 76.98% of the municipal wastewater was collected. In addition, only 4.56 thousand hm^3/year (144.71 m^3/s) were treated, which indicates that 51.7% of the total municipal wastewater generated was treated in that year. Whereas 67.15% of the wastewater that was collected by the sanitation systems was treated [43].

On the other hand, in the same period, sewerage coverage in Mexico reached 95.2%, which means that approximately 119.3 million people, in that year, had access to sewerage services. Therefore, the possibility of reusing municipal wastewater represents a promising alternative source of water supply. However, in the absence of efficient treatment, wastewater can constitute an important source of microbiological risk to human health. Available data on the inventory of WWTPs indicate that from 2004 to 2020 it increased from 394 to 2786 (**Figure 3**). This means that, on average, 85 WWTPs were built each year and the total increase is 316%. This indicates that there is a great effort in the country to increase the number of existing plants and that sanitation plans and policies have had results [43].

2.4.2 Types of wastewater treatment

Wastewater treatment consists of a process to remove contaminants, mainly from domestic wastewater, which includes physical, chemical, and biological processes [46]. WWTPs in Mexico have different types of treatment (**Table 3**). The most used process in WWTPs is activated sludge (39.7%); this type of process can treat wastewater with high organic loads and has the potential to produce biogas, as they

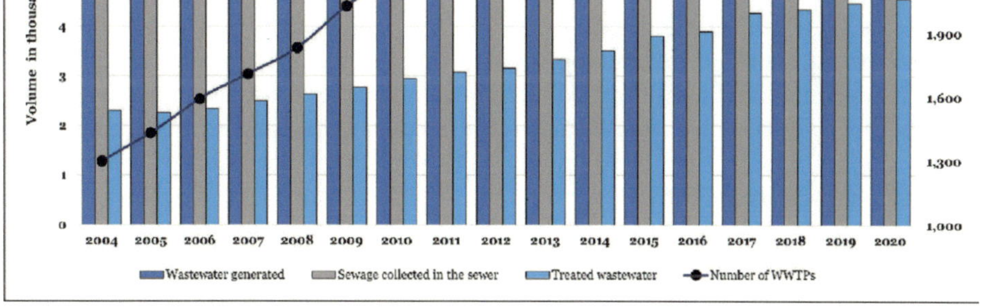

Figure 3.
Number of WWTPs operating in Mexico and treated flow from 2004 to 2020 [43].

Description	Treatment (%)	Discharge volume in m³/s	Volume in thousands of hm³/ year
Activated sludge	39.7	57.45	1.81
Dual	21.9	31.69	1.00
Stabilization lagoons	4.0	5.79	0.18
Aerated lagoons	3.9	5.64	0.18
Advanced primary	3.0	4.34	0.14
Others	1.9	2.75	0.09
UASB*	1.1	1.59	0.05
Biological filters	0.9	1.30	0.04
Primary	0.2	0.29	0.01

**Up-flow Anaerobic Sludge Blanket.*

Table 3.
Main treatment processes and volume of wastewater discharge in Mexico [43].

Stages	Description
Preliminary	Large solids and sand are removed by sieving.
Primary	The water is left to rest so that solids can sink to the bottom and oil and grease can rise to the surface.
Secondary	Biological treatment usually based on the activated sludge process, used to remove dissolved and colloidal compounds from wastewater. The elimination of nutrients and the accumulation of biomass also occurs.
Tertiary	This can be considered as the final stage necessary to achieve the specific quality. In some cases, disinfection processes are carried out by applying chlorine, ozone, or ultraviolet (UV) radiation before the treated wastewater is discharged into aquatic ecosystems or reused.

Table 4.
Stages for wastewater treatment in a WWTP [9, 41, 46, 49].

are considered simple to construct and have low operating costs compared to modern technologies. Modern technologies include membrane processes (reverse osmosis and nanofiltration), membrane bioreactors, advanced oxidation processes, ozonation, photocatalysis, and radiation, which are becoming attractive approaches for WWTPs despite their high maintenance and operation costs [44].

In this sense, conventional WWTPs also need significant financial investments to improve the facilities and processes in each of their stages, and this situation demands the implementation of new technologies [47]. Nevertheless, current conventional methods are considered a widely used technology, capable of producing a safe effluent to protect ecosystems and human health [48]. According to Hong et al. [9], conventional WWTPs operate with the following treatment stages: preliminary, primary, secondary, and tertiary (**Table 4**).

2.4.3 Sanitation system in Acapulco

The city of Acapulco, Guerrero, is a beach tourist resort located in southeastern Mexico at 16°56′56″ N and 99°55′12″ [50]. Its main urban area is developed around

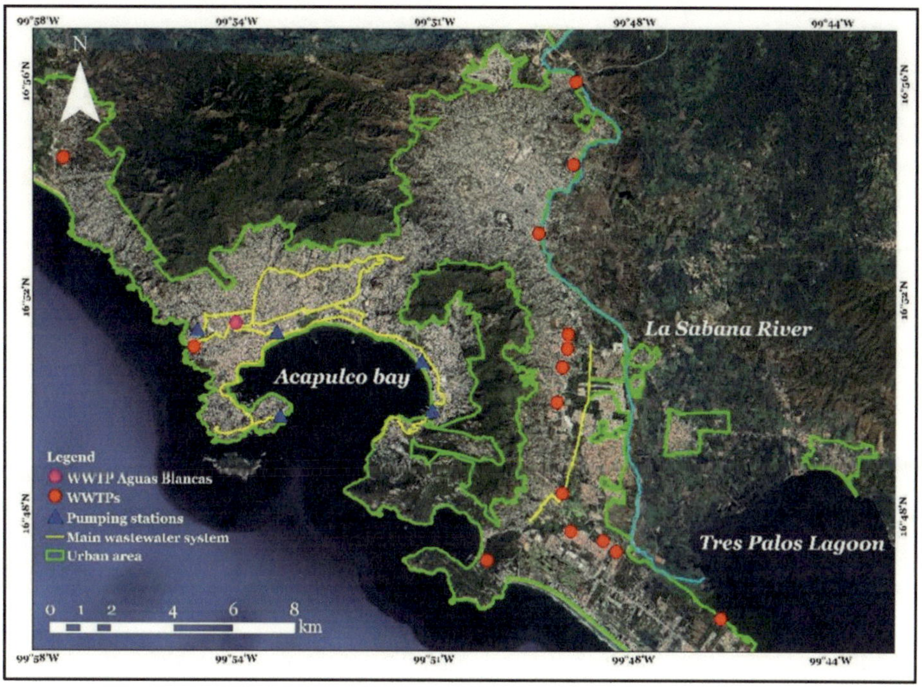

Figure 4.
Location of WWTPs in the municipality of Acapulco [6, 34, 50]

a semicircular bay characterized by a rugged topography (**Figure 4**). The climate is of the Aw1 warm sub-humid type with an average annual temperature of 27.8°C and an average annual rainfall of 561 mm [51]. The accelerated urban development that the city has experienced in recent decades and the deficiencies in public services have caused serious water and soil contamination problems, leading to social and environmental vulnerability [52].

Acapulco Bay and the La Sabana River are exposed to contamination due to wastewater discharges and poor solid waste management [53, 54]. To treat wastewater, the city has 18 WWTPs consisting of conventional treatment (activated sludge), managed by the local water agency. Wastewater is collected and transported through the sewerage network, directing the raw water to the treatment units, which are distributed in different parts of the city, discharging the treated wastewater into the La Sabana River or the Pacific Ocean coast [6]. As a study case, the "Aguas Blancas" WWTP was considered as one of the most representative of the city due to its level of efficiency, treatment capacity, and quality of treated water (**Table 5**).

3. Potential reuse of treated wastewater in Mexico

3.1 Reuse of wastewater from environmental, social, economic, political, and technical perspectives

Currently, wastewater treatment has two main purposes: sanitation and reuse [55], the first is related to human health and environmental protection, and the second to

Features	WWTP Aguas Blancas
Location	16.860015°–99.908553°
Process	Activated sludge
Type of treatment	Extended aeration
Type of wastewater received	Domestic
Capacity in liters per second (L/s)	1350 L/s
Target population	600,000 inhabitants
Sludge treatment	Centrifuge
Receiving body	Pacific Ocean coast
Tertiary treatment	UV rays
Efficiency level	90%
Legal compliance	NOM-001-SEMARNAT-1996

Source: Information supplemented by Herrera-Navarrete et al. [6].

Table 5.
Characteristics of the main WWTP in Acapulco [6].

mitigate contamination and scarcity problems. The reuse of wastewater is not a new practice; there are indications from ancient civilizations, where it was used for crop irrigation [16]. Among the wide variety of applications of treated wastewater are irrigation, groundwater recharge, domestic use, industrial applications, and even the production of drinking water with high-tech treatments [56].

Table 6 shows the various reuses of treated wastewater using different degrees of purification, but it is also reused untreated, especially in underdeveloped countries in Latin America, Asia, and Africa, with water scarcity arises the need for high-quality effluents. Therefore, it is necessary to promote conventional and advanced tertiary treatments for wastewater treatment [57].

Globally, the largest demand for water comes from the agricultural sector, which accounts for approximately 70% of all freshwater withdrawals; therefore, the reuse of treated wastewater for this sector has become one of the most reliable and low-cost alternatives [58]. According to Valdes et al. [2] in several studies, the reuse of treated wastewater has been demonstrated; however, this practice reveals some advantages and disadvantages (**Table** 7). Therefore, the level of treatment for reuse depends on the water quality requirements for the intended use [59]. In this sense, it is important to determine the environmental, social, economic, political, and technical aspects involved in the reuse of treated wastewater [60].

3.2 Environmental aspect

WWTPs aim to protect water resources and human health by reducing nutrients and pathogens discharged to water bodies. However, global problems such as pollution and water scarcity induce toward reuse, an issue of high relevance [44]. The reuse of wastewater can be a viable alternative to solve problems mainly related to scarcity, but it is also important not to lose sight of the issue of ECs; WWTPs have been identified as major sources of this type of pollutants affecting aquatic ecosystems [61].

Reuse	Description
Municipal	Irrigation of public parks, sports facilities, gardens, parkways; street cleaning; fire protection systems; vehicle washing and toilet flushing.
Agricultural	Irrigation of processed and unprocessed food crops, animal pastures, fodder, fiber, seed crops, ornamental flowers, orchards, hydroponic crops, aquaculture, greenhouses, viticulture, etc.
Industrial	Cooling water, cooling towers, washing water, concrete manufacturing, soil compaction, and dust control.
Recreational	Irrigation of golf courses, recreational reservoirs with or without public access (e.g. for fishing and navigation), and ornamental reservoirs without public access and artificial snow.
Environmental	Recharge of aquifers, wetlands, and swamps; ecological flows, wildlife habitat, and forestry.
Drinking use	Aquifer recharge and advanced treatment to obtain the quality of drinking water.

Table 6.
Possible reuse of treated wastewater on a global scale [57].

Advantages	Disadvantages
• Reduces stress from water sources.	• High investment.
• Minimizes freshwater pollution.	• Distance to transport water.
• Source of nutrients, useful as a fertilizer.	• It requires advanced technologies.
• Greater availability of freshwater resources.	• Social acceptance.
• Increased water availability for urban areas.	• It causes the accumulation of chemical and biological contaminants in the soil.
• Affordable prices for treated water.	• Emerging contaminants could enter the food chain.

Table 7.
Advantages and disadvantages of reuse of treated wastewater [2].

3.3 Social aspect

One of the main challenges in the reuse of treated wastewater is the degree of acceptance, influenced by many factors: education, risk awareness, degree of water scarcity or availability of alternative water sources, calculated costs and benefits, trust and knowledge, issues of choice, environmental attitudes, and participation in decision-making, in addition to other cultural, religious, and socioeconomic factors [60].

3.4 Economic aspect

Water reuse for industrial or irrigation purposes is considered to have a lower environmental impact and cost compared to other alternative water supplies such as: water transfers or desalination; however, these practices are carried out in a limited way due to legal and social issues [62]. On the other hand, considerable investment is required to renovate and upgrade WWTPs to operate more efficiently [3]. The cost of water reclamation (including all costs, investment, operation, and maintenance) from wastewater to the level of drinking water which ranges internationally between 0.70 and 1 USD/m^3 [63, 64].

3.5 Political aspect

Decision-making is strongly based on political interests and social pressure. Alignment of common objectives on public health, environmental protection, and agricultural development between local authorities and different sectors is needed to overcome these challenges. On the other hand, they point out that the short terms of the municipal government affect the follow-up of a long-term improvement plan for the supply and sanitation of water, postponing financial resources with state or federal instances [60].

3.6 Technical aspect

Most WWTPs are not designed to handle the excess volumes of heavy rainfall, thus affecting the hydraulic systems and causing wastewater overflows. Another technical aspect refers to the managers of the water systems, since they do not have basic training in issues related to laws and regulations with water resources management or knowledge of urban hydraulic infrastructure [60].

3.7 Regulatory framework

The progress of planned wastewater reclamation and reuse depends not only on technological advances but also on the existence of a robust legal framework that establishes guidelines for reuse that does not entail risks for the beneficiaries. Legislation on wastewater reuse on a global scale is a complicated issue because, while there are countries with legal regulations, others only offer recommendations, each with its own parameters and indicators [65].

In Mexico, water is considered a public resource and is administered by the National Water Commission (CONAGUA) through the National Water Law, which is derived from the Mexican Political Constitution and is embodied in two regulatory articles of great interest, in addition to containing other mandatory regulations (**Table 8**). On the other hand, Mexico is a pioneer in establishing a regulation that describes the requirements for aquifer recharge with treated wastewater [68]. There is a strengthened legal framework; however, the lack of compliance with quality standards in water reuse, mainly in effluents, is noticeable. It is possible that strict legislation could lead to unsafe reuse due to the high costs involved in treatment and monitoring [45].

3.8 Proposal for the reuse of treated water: Acapulco, study case

According to the studies carried out by Martínez-Orgániz et al. [34] and Martínez-Orgániz et al. [69], the "Aguas Blancas" treatment plant complies with the requirements established according to Mexican Standards. The research proposes a modification in the treatment process to avoid the presence of various types of emerging contaminants, and microorganisms such as *E. coli*. It is important to note that this treatment unit discharges its effluent into the sea. Due to its level of treatment in recent years, it is considered one of the best plants nationwide; however, its efficiency potential can be improved under a design similar to the Orange Country plant (**Figure 5**), which is considered a model plant worldwide [15].

The city of Acapulco is one of the most important beach resorts in the country. The current economic situation of the municipality and of the Municipal Drinking

Year*	Legal framework	Description
1917	Political Constitution	Article 27 establishes guidelines based on the types of water bodies. Whereas Article 115 establishes that municipal governments are in charge and responsible for the treatment and disposal of wastewater in Mexico.
1971	Federal Law to Prevent and Control Environmental Contamination (FLPCEC).	First environmental law to decree natural areas protected from contamination and overexploitation.
1982	Federal Law on Environmental Protection.	Law that replaces the FLPCEC from an environmental policy approach, ans it was broadened and oriented toward prevention and health.
1988	General Law of Ecological Balance and Environmental Protection	Article 117 establishes that "wastewater of urban origin must receive treatment prior to its discharge into rivers, basins, vessels, marine waters and other deposits or streams of water, including subsoil waters."
1991	Federal Law of Duties (Ministry of Finance and Public Credit)	It contains the obligation to pay duties for discharges of wastewater into national waters, such as riverbeds, lakes, or seas.
1992	National Waters Law	It declares of public utility the installation of wastewater treatment plants and the execution of measures for the reuse of such waters, as well as the construction of works for the prevention and control of water contamination.
1995	NOM-067-ECOL-1994	It defines the general parameters of municipal wastewater discharges.
1996	NOM-001-SEMARNAT-1996	Establishes the maximum permissible limits of pollutants in wastewater discharges into national waters and property.
1998	NOM-002-SEMARNAT-1996	It establishes the maximum permissible limits of pollutants in wastewater discharges in municipal sewage systems.
1998	NOM-003-SEMARNAT-1996	It establishes the maximum permissible limits of pollutants for treated wastewater reused in services to the public.
2003	NOM-004-SEMARNAT-2002	It establishes the specifications and maximum permissible limits of contaminants in sludge and biosolids for their use and final disposal.
2009	NOM-014-CONAGUA-2003	It establishes the requirements for the artificial recharge of aquifers with treated wastewater.
2022	NOM-001-SEMARNAT-2021	It repeals the 1996 standard and updates some parameters. Effective April 2023.

**Year of publication in the Diario Oficial de la Federación (DOF).*

Table 8.
Evolution of the legal framework related to wastewater [18, 66, 67].

Water and Sewerage Operation Agency (CAPAMA) does not allow for the implementation of treatment systems that involve large-scale restructuring. On the other hand, the Aguas Blancas WWTP is located within the urban area of Acapulco; an area with very few possibilities for expansion which limits the creation of new facilities containing other processes (e.g. Microfiltration, Reverse Osmosis, Granular Activated Carbon).

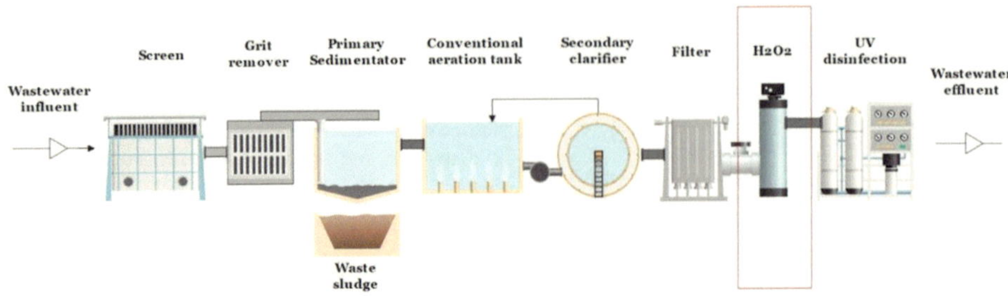

Figure 5.
Outline of the process as a proposal for the Aguas Blancas WWTP (adapted from Orange Country [15]).

Of the Advanced Oxidation Processes (AOP), Ozonation and Hydrogen Peroxide (H_2O_2) combined with UV (Orange Country) is the most efficient technology for the generation of hydroxyl radicals (OH^-).

In addition, hydrogen peroxide is considered a green oxidant, as it decomposes into water and oxygen, and its use in wastewater treatment has increased globally in recent years. Therefore, H_2O_2/UV is the best AOP in terms of meeting the technical, economic, and environmental constraints for advanced treatment [70]. The proposal is to modify the current treatment process to an Advanced Oxidation Treatment (AOP) using hydrogen peroxide (H_2O_2). Prior to additional UV treatment, it is suggested to consider lengthening the retention time, which can significantly reduce the discharge of several detected ECs and possible adverse effects to the environment and public health [70]. However, the implementation of this proposal could result in the potential reuse of treated water for agriculture, landscaping, and aquifer recharge, which are of great relevance for the development of the region.

3.8.1 Agriculture

Agriculture consumes between 50% and 90% of the total water demand; therefore, wastewater reuse for this sector is considered a solution to overcome global water stress [71]. Valdes et al. [2] state that reuse of treated wastewater involves benefits and risks as demonstrated in many studies. However, the limitations are reduced with improved treatment technology, which increases the reliability of treated wastewater production and meet the standards. Consequently, many countries have succeeded in treating wastewater to an acceptable quality for unrestricted reuse in agriculture [72].

3.8.2 Landscape

The reuse of treated wastewater for landscape irrigation, particularly on a regional scale, is an attractive option, since it can combine essential basic needs such as pollution control, preservation of water quality, and volume for sustainable irrigation of green areas; therefore, the reuse of treated wastewater in this area can be considered a viable alternative for irrigation of parks, sports areas, schoolyards, green areas of residential settlements, and golf courses [45, 59]. Landscape irrigation requires large volumes of freshwater; according to the World Health Organization (WHO), urban green areas are an important element for improving the quality of life [2].

3.8.3 Aquifer recharge

According to Seguí et al. [15], treated wastewater and rainwater represent a great opportunity for their use in the recharge of aquifers; however, to achieve this strategy, hydraulic infrastructure in optimal conditions is required, since they are mixed in the municipal sewerage systems. On the other hand, several studies assure a low public health risk caused by water contamination in the action that involves the recharge of aquifers with reclaimed water from WWTPs, a concern that leads to consider the application of a regulation in Mexico to guarantee the quality of treated water. In this sense, the recharge of an aquifer with reclaimed water is presented as the best environmental contribution to counteract water scarcity. Other benefits include the increase in groundwater reserves and the preservation of aquatic ecosystems [15].

4. Conclusions

The reuse of treated wastewater should be considered a common practice. It is convenient to establish effective social communication so that users are informed about the environmental benefits that this practice entails in their daily lives, and on the positive impact that its reuse would have on agriculture, a sector that demands large quantities of water. On the other hand, the optimal operational functioning of WWTPs must be a priority for local governments, which implies providing them with technical and financial resources, in addition to guaranteeing that treated wastewater effectively reaches the WWTPs through an effective collection system.

In this sense, the environmental legal framework related to water quality and mainly treated wastewater must be strictly and continuously monitored, as failure to comply not only implies sanctions, but also produces negative consequences for the environment. In order to strengthen current regulations, CEs must be taken into account, since they put public health at risk and cause adverse effects on aquatic eco systems. Finally, this study proposes an adjustment to a WWTP in the municipality of Acapulco in its water treatment line, which consists of complementing the system with an advanced process (AOP) using hydrogen peroxide (H_2O_2) before UV disinfection. This improvement eliminates a wide variety of ECs and guarantees water quality for potential reuse in agriculture, in the irrigation of landscape or urban green areas, but especially for aquifer recharge. All these practices are of great relevance mainly for those that demand a greater volume of water.

Conflict of interest

The authors declare no conflict of interest.

Author details

Ángeles Martínez-Orgániz[1], Ricardo Herrera-Navarrete[2*], Daniel Pineda-Mora[2], Viridiana Del Carmen-Niño[3] and Rosa Iris Balbuena-Hernández[4]

1 Autonomous University of Guerrero, School of Natural Sciences, Carretera Nacional Chilpancingo-Petaquillas, Chilpancingo, Guerrero, Mexico

2 Autonomous University of Guerrero, Regional Development Science Center, Acapulco, Guerrero, Mexico

3 Autonomous University of Guerrero, School of Sustainable Development, Tecpan de Galeana, Guerrero, Mexico

4 Faculty of Nursing, Autonomous University of Guerrero, Acapulco, Guerrero, Mexico

*Address all correspondence to: rherrera@uagro.mx

References

[1] Mishra B, Kumar P, Saraswat C, Chakraborty S, Gautam A. Water security in a changing environment: Concept, challenges and solutions. Water. 2021;**13**:1-21

[2] Valdes A, Aguilera EN, Tobón G, Samaniego L, Díaz L, Carlos S. Potential uses of treated municipal wastewater in a semiarid region of Mexico. Sustainability. 2019;**11**:1-24

[3] Romano O, Akhmouch A. Water governance in cities: Current trends and future challenges. Water. 2019;**11**:1-9

[4] Abd-Elhamid HF, Abd-Elmoneem SM, Abdelaal GM, Zelenakova M, Vranayova Z, Abd-Elaty I. Investigating and managing the impact of using untreated wastewater for irrigation on the groundwater quality in arid and semi-arid regions. International Journal of Environmental Research and Public Health. 2021;**18**(14):1-17

[5] Wang D, Hubacek K, Shan Y, Gerbens-Leenes W, Liu J. A review of water stress and water footprint accounting. Water. 2021;**13**:1-15

[6] Herrera-Navarrete R, Colin-Cruz A, Arellano-Wences HJ, Sampedro-Rosas ML, Rosas-Acevedo JL, Rodriguez-Herrera AL. Municipal wastewater treatment plants: Gap, challenges, and opportunities in environmental management. Environmental Management. 2022;**69**:75-88. DOI: 10.1007/s00267-021-01562-y

[7] Vasilachi I, Asiminicesei D, Fertu D, Gavrilescu M. Occurrence and fate of emerging pollutants in water environment and options for their removal. Water. 2021;**13**:1-34

[8] Zhang L, Shen Z, Fang W, Gao G. Composition of bacterial communities in municipal wastewater treatment plant. Science Total Environment. 2019;**689**:1181-1191. DOI: 10.1016/j.scitotenv.2019.06.432

[9] Hong E, Yeneneh AM, Sen TK, Ang HM, Kayaalp A. A comprehensive review on rheological studies of sludge from various sections of municipal wastewater treatment plants for enhancement of process performance. Advances in Colloid and Interface Science. 2018;**257**:19-30. DOI: 10.1016/j.cis.2018.06.002

[10] Yang L, Wen Q, Chen Z, Duan R, Yang P. Impacts of advanced treatment processes on elimination of antibiotic resistance genes in a municipal wastewater treatment plant. Frontiers of Environmental Science & Engineering. 2019;**13**:1-10. DOI: 10.1007/s11783-019-1116-5

[11] Elizondo LS, Mendoza-Espinosa LG. An analysis of water scarcity in a drought prone city: The case of Ensenada, Baja California, Mexico. Tecnología y ciencias del agua. 2020;**11**:1-55. DOI: 10.24850/j-tyca-2020-02-01

[12] Pinos JA, Malo AJ. El derecho humano de acceso al agua: una revisión desde el Foro Mundial del Agua y la gestión de los recursos hídricos en Latinoamérica. INVURNUS. 2018;**13**:12-20

[13] Milena A, Ortega Y, Mujica E. Potencial de reutilización del efluente de la planta de tratamiento de agua residuales de Timaná- Huila para riego de pasto estrella (Cynodon Plectostachius). Ingenieria y Región. 2017;**2017**:33-44

[14] Abeledo-Lameiro MJ, Ares-Mazas E, Gomez-Couso H. Use of ultrasound irradiation to inactivate Cryptosporidium parvum oocysts in effluents from municipal wastewater treatment plants. Ultrasonics Sonochemistry. 2018;**48**:118-126. DOI: 10.1016/j.ultsonch.2018.05.013

[15] Seguí LA, Moeller-Chávez G, De Andrés A. Mexico, the water stress: Challenges and opportunities in wastewater treatment and reuse. Water Policy in Mexico. 2019;**2019**:75-87

[16] Ghafoori S, Hassanpour H, Mohamadvali H, Taherei P. Enhancing the method of decentralized multi-purpose reuse of wastewater in urban area. Sustainability. 2021;**13**:1-12

[17] De Anda J. Saneamiento descentralizado y reutilización sustentable de las aguas residuales municipales en México. Sociedad y Ambiente. 2017;**14**:119-143

[18] CONAGUA. Situación del subsector Agua Potable, Alcantarillado y Saneamiento. Comisión Nacional del Agua. México: Secretaría de Medio Ambiente y Recursos Naturales; 2021

[19] Martínez A, Hernández-Flores G, Toribio-Jiménez J, Becerril JE, Sampedro ML, Mendoza-Ramos JE. Tecnologías para la remoción de antibióticos presentes en agua residual. In Tópicos sobre contaminantes y contaminación del agua, Editorial, C., Ed. México; 2019. pp. 419-440

[20] Suárez A, Lembo C, Ríos JY, Vieitez D, Astesiano G, Corzo JF. Casos de estudio en Asociaciones Público-Privadas en América Latina y el Caribe. In: Planta de tratamiento de aguas residuales Atotonilco. México; 2019

[21] Barceló LD, López MJ. Contaminación y calidad química del agua: el problema de los contaminantes emergentes. Fundación Nueva Cultura del Agua. Panel científico-técnico de seguimiento de la política de aguas. Convenio Universidad de Sevilla-Ministerio de Medio Ambiente; 2008. pp. 1-27

[22] Ding X, Wang S, Jiang G, Huang G. A simulation program on change trend of pollutant concentration under water pollution accidents and its application in Heshangshan drinking water source area. Journal of Cleaner Production. 2017;**167**:326-336. DOI: 10.1016/j.jclepro.2017.08.094

[23] Naddeo V, Secondes MFN, Borea L, Hasan SW, Ballesteros F Jr, Belgiorno V. Removal of contaminants of emerging concern from real wastewater by an innovative hybrid membrane process - UltraSound, adsorption, and membrane ultrafiltration (USAMe(R)). Ultrasonics Sonochemistry. 2020;**68**:105237. DOI: 10.1016/j.ultsonch.2020.105237

[24] Adeleye AS, Xue J, Zhao Y, Taylor AA, Zenobio JE, Sun Y, et al. Abundance, fate, and effects of pharmaceuticals and personal care products in aquatic environments. Journal of Hazardous Materials. 2022;**424**:127284. DOI: 10.1016/j.jhazmat.2021.127284

[25] Le-Minh N, Khan SJ, Drewes JE, Stuetz RM. Fate of antibiotics during municipal water recycling treatment processes. Water Research. 2010;**44**:4295-4323. DOI: 10.1016/j.watres.2010.06.020

[26] Watkinson AJ, Murby EJ, Costanzo SD. Removal of antibiotics in conventional and advanced wastewater treatment: Implications for environmental discharge and wastewater recycling. Water Research. 2007;**41**:4164-4176. DOI: 10.1016/j.watres.2007.04.005

[27] Arpin-Pont L, Bueno MJ, Gomez E, Fenet H. Occurrence of PPCPs in the marine environment: A review. Environmental Science and Pollution Research International. 2016;**23**:4978-4991. DOI: 10.1007/s11356-014-3617-x

[28] Hilton MJ, Thomas KV. Determination of selected human pharmaceutical compounds in effluent and surface water samples by high-performance liquid chromatography–electrospray tandem mass spectrometry. Journal of Chromatography A. 2003;**1015**:129-141. DOI: 10.1016/s0021-9673(03)01213-5

[29] McGrane SJ. Impacts of urbanisation on hydrological and water quality dynamics, and urban water management: A review. Hydrological Sciences Journal. 2016;**61**:2295-2311. DOI: 10.1080/02626667.2015.1128084

[30] Wang J, Tian Z, Huo Y, Yang M, Zheng X, Zhang Y. Monitoring of 943 organic micropollutants in wastewater from municipal wastewater treatment plants with secondary and advanced treatment processes. Journal of Environmental Sciences (China). 2018;**67**:309-317. DOI: 10.1016/j.jes.2017.09.014

[31] Estrada-Arriaga EB, Cortes-Munoz JE, Gonzalez-Herrera A, Calderon-Molgora CG, de Lourdes R-HM, Ramirez-Camperos E, et al. Assessment of full-scale biological nutrient removal systems upgraded with physico-chemical processes for the removal of emerging pollutants present in wastewaters from Mexico. Science Total Environment. 2016;**571**:1172-1182. DOI: 10.1016/j.scitotenv.2016.07.118

[32] Hernández-Tenorio R, Guzmán-Mar JL, Hinojosa-Reyes L, Ramos-DelgadoN, Hernández-RamírezA. Determination of pharmaceuticals discharged in wastewater from a public hospital using LC-MS/MS technique. Journal of the Mexican Chemical Society. 2021;**65**. DOI: 10.29356/jmcs.v65i1.1439

[33] Lopez-Velazquez K, Guzman-Mar JL, Saldarriaga-Norena HA, Murillo-Tovar MA, Hinojosa-Reyes L, Villanueva-Rodriguez M. Occurrence and seasonal distribution of five selected endocrine-disrupting compounds in wastewater treatment plants of the metropolitan area of Monterrey, Mexico: The role of water quality parameters. Environmental Pollution. 2021;**269**:116223. DOI: 10.1016/j.envpol.2020.116223

[34] Martínez-Orgániz Á, Bravo JE, Llompart M, Dagnac T, Pablo J, Vázquez L, et al. Emerging pollutants and antibiotics removed by conventional activated sludge followed by ultraviolet radiation in a municipal wastewater treatment plant in Mexico. Water Quality Research Journal. 2021;**56**:167-179. DOI: 10.2166/wqrj.2021.013

[35] Rivera-Jaimes JA, Postigo C, Melgoza-Aleman RM, Acena J, Barcelo D, Lopez de Alda M. Study of pharmaceuticals in surface and wastewater from Cuernavaca, Morelos, Mexico: Occurrence and environmental risk assessment. Science Total Environment. 2018;**613-614**:1263-1274. DOI: 10.1016/j.scitotenv.2017.09.134

[36] Robledo VH, Velázquez MA, Montañez JL, Pimentel JL, Vallejo AA, López MD, et al. HidroquÍmica Y Contaminantes Emergentes En Aguas Residuales Urbano Industriales De Morelia, MichoacÁn MÉxico. Revista Internacional de Contaminación Ambiental. 2017;**33**:221-235. DOI: 10.20937/rica.2017.33.02.04

[37] Navarro Frómeta AE, Leyva ZC, Mendoza JC. Tópicos sobre

contaminantes y contaminación del agua. Primera Edición: Editorial CLAVE; 2019

[38] WWAP. The United Nations World Water Development Report 2017. Wastewater: The Untapped Resource. United Nations World Water Assessment Programme (WWAP). Paris: UNESCO; 2017

[39] LAN. Ley de Aguas Nacionales. Cámara de diptutados del H. Congreso de la Unión. Última Reforma del Diario Oficial de la Federación (DOF) 06-01-2020. México. 2020

[40] Almeida P, Albuquerque T, Antunes M, Ferreira A, Pelletier G. Effects of wastewater treatment Plant's discharges on a freshwater ecosystem—A case study on the Ramalhoso River (Portugal). Water, Air, & Soil Pollution. 2021;**232**(5):1-11. DOI: 10.1007/s11270-021-05131-1

[41] Noyola A, Morgan-Sagastume JM, Guereca LP. Selección de Tecnologías para el Tratamiento de Aguas Residuales Municipales: Guía de apoyo para ciudades pequeñas y medianas. Universidad Nacional Autónoma de México. México: Primera Edición; 2013

[42] Pulido-Madrigal L. Riego con aguas residuales para depuración de contaminantes. H2 O Gestión del agua. 2017;**14**:12-17

[43] CONAGUA. Comisión Nacional del Agua - Sistema Nacional de Información del Agua (SINA). [Internet]. 2020. Availabe online: http://sina.conagua.gob.mx/sina/ [Accessed: June 02, 2022]

[44] Tang J, Zhang C, Shi X, Sun J, Cunningham JA. Municipal wastewater treatment plants coupled with electrochemical, biological and bio-electrochemical technologies: Opportunities and challenge toward energy self-sufficiency. Journal of Environmental Management. 2019;**234**:396-403. DOI: 10.1016/j.jenvman.2018.12.097

[45] Garcia D, Muñoz G, Arteaga A, Ojeda-Revah L, Mladenov N. Greening urban areas with decentralized wastewater treatment and reuse: A case study of Ecoparque in Tijuana, Mexico. Water. 2022;**14**:1-18

[46] Demirbas A, Edris G, Alalayah WM. Sludge production from municipal wastewater treatment in sewage treatment plant. Energy Sources, Part A: Recovery, Utilization, and Environmental Effects. 2017;**39**:999-1006. DOI: 10.1080/15567036.2017.1283551

[47] Szymanski K, Morawski AW, Mozia S. Effectiveness of treatment of secondary effluent from a municipal wastewater treatment plant in a photocatalytic membrane reactor and hybrid UV/H2O2 – Ultrafiltration system. Chemical Engineering and Processing - Process Intensification. 2018;**125**:318-324. DOI: 10.1016/j.cep.2017.11.015

[48] Papa M, Foladori P, Guglielmi L, Bertanza G. How far are we from closing the loop of sewage resource recovery? A real picture of municipal wastewater treatment plants in Italy. Journal of Environmental Management. 2017;**198**:9-15. DOI: 10.1016/j.jenvman.2017.04.061

[49] Caicedo C, Rosenwinkel KH, Exner M, Verstraete W, Suchenwirth R, Hartemann P, et al. Legionella occurrence in municipal and industrial wastewater treatment plants and risks of reclaimed wastewater reuse: Review. Water Research. 2019;**149**:21-34. DOI: 10.1016/j.watres.2018.10.080

[50] INEGI. Aspectos Geográficos de Guerrero, México. Instituto Nacional

de Estadística y Geografía. [Internet]. 2018. Availabe online: https://en.www.inegi.org.mx/contenidos/app/areasgeograficas/resumen/resumen_12.pdf [Accessed: June 10, 2022]

[51] SMN. Precipitación (mm) por Entidad Federativa y Nacional. Servicio Meteorológico Nacional (SMN). Comisión Nacional del Agua. México: Secretaría de Medio Ambiente y Recursos Naturales; 2022

[52] Rodríguez A, Olivier B, López R, Barragán MC, Cañedo R, Valera MÁ. Contaminación y riesgo sanitario en zonas urbanas de la subcuenca del río de La Sabana, ciudad de Acapulco. México. Gestión y Ambiente. 2013;**16**:85-95

[53] IMTA. Estudio de clasificación de la Bahía de Acapulco, Gro. Fondos Sectoriales de Investigación y Desarrollo sobre el Agua Convocatoria 2006-01 de CONAGUA-CONACYT con número de proyecto 48801. Instituto Mexicano de Tecnología del Agua (IMTA). México: Secretaria de medio Ambiente y Recursos Natruales (SEMARNAT); 2009

[54] Pineda D, Toribio J, Leal MT, Juarez AL, González J, Batista RA. Emerging water quality issues along Rio de la Sabana, Mexico. Journal of Water Resource and Protection. 2018;**10**:621-636

[55] Salgot M, Folch M. Wastewater treatment and water reuse. Current Opinion in Environmental Science & Health. 2018;**2**:64-74

[56] Collivignarelli MC, Abba A, Miino MC, Caccamo FM, Torretta V, Rada EC, et al. Disinfection of wastewater by UV-based treatment for reuse in a circular economy perspective. Where are we at? International Journal of Environmental Research and Public Health. 2020;**18**:1-24

[57] Prats-Rico D. La reutilización de aguas depuradas regeneradas a escala mundial: análisis y prospectivas. Agua y Territorio. 2016;**2016**:10-21. DOI: 10.17561/at.v0i8.3292

[58] AbdelMoula S, Sorour MT, Aly SAA. Cost analysis and health risk assessment of wastewater reuse from secondary and tertiary wastewater treatment plants. Sustainability. 2021;**13**(23):1-17

[59] Diogo AF, Resende RA, Oliveira AL. Optimised selection of water supply and irrigation sources—A case study on surface and underground water, desalination, and wastewater reuse in a Sahelian coastal arid region. Sustainability. 2021;**13**:1-24

[60] De Anda J, Shear H. Sustainable wastewater management to reduce freshwater contamination and water depletion in Mexico. Water. 2021;**13**:1-19

[61] Tran NH, Reinhard M, Gin KY. Occurrence and fate of emerging contaminants in municipal wastewater treatment plants from different geographical regions-a review. Water Research. 2018;**133**:182-207. DOI: 10.1016/j.watres.2017.12.029

[62] Rizzo L, Krätke R, Linders J, Scott M, Vighi M, de Voogt P. Proposed EU minimum quality requirements for water reuse in agricultural irrigation and aquifer recharge: SCHEER scientific advice. Current Opinion in Environmental Science & Health. 2018;**2**:7-11. DOI: 10.1016/j.coesh.2017.12.004

[63] Seguí-Amórtegui L, Alfranca-Burriel O, Moeller-Chávez G. Metodología para el análisis técnico-económico de los sistemas de regeneración y reutilización de las aguas residuales. Tecnología y ciencias del agua. 2014;**55**:55-70

[64] Tello P, Mijailova P, Chamy R. Uso seguro del agua para el reúso. México: Programa hidrologico Internacional; 2016

[65] Asano T, Cotruvo JA. Groundwater recharge with reclaimed municipal wastewater: Health and regulatory considerations. Water Research. 2004;**38**:1941-1951. DOI: 10.1016/j.watres.2004.01.023

[66] Anglés M, Rovalo M, Tejado M. Manual de derecho ambiental mexicano. Mexico: Universidad Autónoma de México. Instituti de investigaciones juridicas; 2021

[67] Maya JM, Pineda N. Avances, estancamiento y limitaciones de la política de saneamiento en México 1998-2014. Entreciencias: Diálogos en la Sociedad del Conocimiento. 2018;**6**:35-50

[68] Gilabert-Alarcón C, Salgado-Méndez S, Daesslé L, Mendoza-Espinosa L, Villada-Canela M. Regulatory challenges for the use of reclaimed water in Mexico: A case study in Baja California. Water. 2018;**10**:1-22

[69] Martínez-Orgániz A, Garza-Ramos U, Sampedro-Rosas ML, González-González J, Nava-Faustino G, Toribio J. Patotipos y resistencia a antibióticos de Escherichia Coli en agua residual. Revista Internacional de Contaminación Ambiental. 2020;**36**(4):957-966. DOI: 10.20937/rica.53711

[70] Chong MN, Sharma AK, Burn S, Saint CP. Feasibility study on the application of advanced oxidation technologies for decentralised wastewater treatment. Journal of Cleaner Production. 2012;**35**:230-238. DOI: 10.1016/j.jclepro.2012.06.003

[71] Marinelli E, Radini S, Akyol Ç, Sgroi M, Eusebi AL, Bischetti GB, et al. Water-energy-food-climate Nexus in an integrated peri-urban wastewater treatment and reuse system: From theory to practice. Sustainability. 2021;**13**:1-13

[72] Mizyed N, Mays DC. Reuse of treated wastewater: From technical innovation to legitimization. In: World Environmental and Water Resources Congress 2020; Henderson, Nevada. 2020

Chapter 3

Role of Activated Carbon in Water Treatment

Muthaian Jaya Rajan and Clastin Indira Anish

Abstract

Heavy metals, such as lead, mercury, zinc, aluminum, arsenic, nickel, chromium, and cobalt, are the common pollutants present within the environment from various natural and Industrial sources. Synthetic dyes are commonly used for dyeing and printing in a variety of industries. The traditional methods for the removal of heavy metals and dyes from wastewater are chemical precipitation, ion exchange, adsorption, membrane processes, and evaporation which require high capital investment and running costs. Activated carbon prepared from agricultural wastes and its by-products are good alternative sources for adsorption because they are low-cost, renewable sources with high carbon, volatile contents, low ash, and reasonable hardness. The preparation means of activated carbon are physical and chemical methods. The important advantages of chemical activation over physical activation are the process that can be accomplished even at lower temperatures and the yield obtained in chemical activation tends to be greater since burn-off char can be avoided. In this chapter, the removal of heavy metals and dyes, using activated carbon, which was prepared by using agricultural waste, biomass was presented. This helps the researchers to accumulate knowledge.

Keywords: activated carbon, biomass, adsorption, activation techniques, hardness

1. Introduction

Disposal of dye effluents from various industries containing heavy metals to water bodies causes water pollution. Due to the scarcity of water recycling, wastewater has become a worldwide concern for the past few decades. It is well known that heavy metals in water are harmful and cause toxic effects to human beings when it is consumed and affects the environment. Dyes are the major cause of water pollutants arising from dye manufacturing and textile industries. The waste chemicals and dye-house effluents liberated from industries must be treated properly to minimize the effects on the environment. Many traditional methods of separation, such as physical and chemical treatment, including coagulation, adsorption, filtration, precipitation, electrodialysis, oxidation, and membrane separation, have been used for the treatment of dye-containing effluents. The adsorption process is one of the best effective and cheaper methods of removing pollutants from wastewater. Green adsorbents used nowadays are high-cost and rare. Therefore, the adsorption process requires an up-gradation in its limitations. The adsorbent employed in the process should be inexpensive and readily available. Activated carbon prepared from agricultural biomass is an adaptable adsorbent because of its eminent properties, such as large surface area, pore volume, diverse pore structure, extensive adsorption capacity, and a high degree of surface reactivity. Due to large surface area and pore volume of the activated carbon, it can be employed in the removal of color, odor, and taste from water and wastewater. It can also be applicable for the recovery of natural gas and air purification in inhabited spaces, such as chemical industries, and it can act as catalyst and catalyst support material [1, 2]. Activated carbon can be prepared from various carbonous source materials, such as agricultural waste and textile waste. The adsorptive, chemical, structural, and catalytic properties of activated carbon were not only determined by the fundamental nature of the source, but also depends on the method of preparation and conditions used during the process. The preparation of activated carbon from carbonaceous raw material involves a series of processes that has to be done with almost care.

2. Carbonization and activation

Carbonization and activation are the most crucial steps for activated carbon production because these two processes determine the main surface properties and porous structure of the adsorbent. During carbonization, non-carbon and volatile carbon species are removed. An elementary pore structure with a fixed carbon mass is produced. The carbonized material obtained will be an elementary graphitic crystallite with a disordered and poorly developed porous structure. The process is usually achieved at temperatures below 800°C in a gaseous environment without any existing oxidants. The parameters which determine the quality and yield of the carbonized product are rate of heating, final temperature, processing time at the final temperature, and the nature like physical state of the carbonaceous precursor. The activation process increases the pore volume of the material, also enlarges the width of the pores formed during carbonization, and develops new pores of carbonized materials. Due to this, the property of the adsorbent will be increased after the activation process.

2.1 Physical and chemical activation

Physical and chemical activation are traditional processes to improve the properties of the adsorbents. The physical activation is usually carried out at temperatures between 800 and 1000°C. It takes place in the presence of oxidizing gases like steam, carbon dioxide (CO_2), air, or a mixture of these gases [3–6]. Commercially prepared activated carbon uses steam activation due to its cost-efficiency. However, CO_2 activation develops a narrow micropore in the early stage of activation, whereas steam activation widened the initial microporosity from the beginning. After the process, activated carbon obtained will have lower micropore volume and larger meso and macropore volumes.

During chemical activation, carbonization and activation are carried out in a single-step process. The raw materials infused with chemical agents during chemical activation are thermally breakdown in between 300 and 800**°C**. The most commonly used reagents for chemical activation are zinc chloride ($ZnCl_2$), phosphoric acid (H_3PO_4), sulfuric acid (H_2SO_4), and alkaline salts, such as potassium hydroxide (KOH) and sodium hydroxide (NaOH). These reagents serve as oxidants and dehydrating agents so that carbonization and activation can take place simultaneously.

Activated carbons prepared using KOH (aq) and NaOH (aq) activation obtained a surface area of 2000 $m^2 g^{-1}$ have been [7–9]. Chemical activation using H_3PO_4, $ZnCl_2$, or KOH and physical activation using CO_2 can also develop activated carbon with very high surface area and pore volume [9–12]. Activated carbon prepared from corncob waste biomass has a pore volume of 1.533 $cm^3 g^{-1}$ and a surface area of 2844 $m^2 g^{-1}$. It was obtained due to chemical activation with KOH at a KOH/char ratio of 4, followed by 30 min CO_2 gasification. An important merit of using chemical activation is that it can proceed at a lower temperature and takes less time when compared with physical activation. However, the demerit of chemical activation is further treatment or process required for reusing the leftover chemical reagent.

3. Structure of activated carbon

3.1 Porous structure activated carbon

The higher adsorption capability of activated carbon mainly depends on porous characteristics such as surface area, pore size distribution, and pore volume. Up to 15% of ash content is present in activated carbon in the form of mineral matter. The porous structure of activated carbon forms during the carbonization process and it further develops during the activation process. All activated carbons have different porous structures. The pore system of activated carbon differs from one another, and individual pores may vary in shape and size. Activated carbons possess pores from less than a nanometer to thousand nanometers. Pores are classified according to their average width. The distance between the walls of a slit-shaped pore or the radius of a cylindrical pore is an average width. Conventional classification of pore and width is proposed, and it is officially adopted by the International Union of Pure and Applied Chemistry (IUPAC) [13]. The pore type and its width are shown in **Table 1.**

3.2 Crystalline structure of activated carbon

The microcrystalline structure of activated carbon develops during carbonization. Activated carbon structure is entirely different when compared to graphite. The interlayer spacing is different in graphite than in activated carbon. The interlayer spacing of graphite is 0.335 nm, whereas in activated carbon the interlayer spacing is 0.34 to 0.35 nm. Based on the graphitizing ability of activated carbons, they are classified into two types: graphitizing and non-graphitizing carbons. The graphene layers are oriented parallel to each other in graphitizing carbon. The carbon obtained was delicate due to the weak cross-linking between the neighbor micro crystallites and had a less developed porous structure. The non-graphitizing carbons are hard in nature. Strong cross-linking between crystallites in non-graphitizing carbons shows

Type of pore	Width
Micro	<2 nm
Meso	2–50 nm
Macro	>50 nm

Table 1.
Classification of pore.

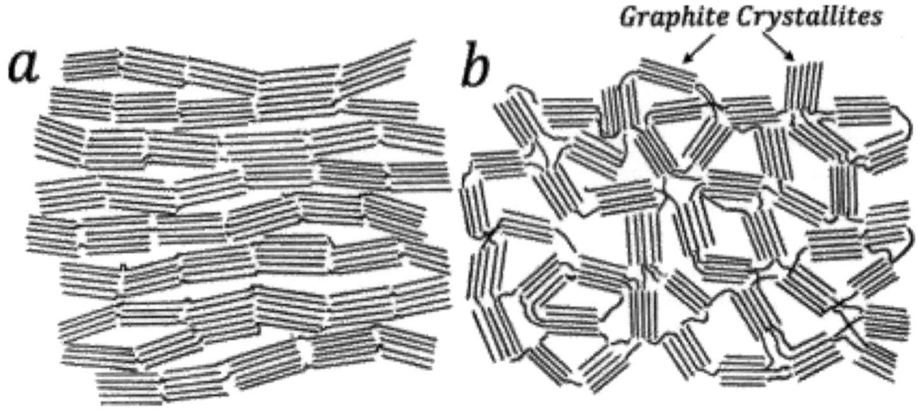

Figure 1.
The structural difference between graphitizing (a) and non-graphitizing (b) carbons [14].

well-developed micropores structure. The formations of non-graphitizing structures with strong crosslink's are promoted by the presence of associated oxygen or by the insufficiency of hydrogen in the original raw material. The structural differences between graphitizing and non-graphitizing carbons are shown in **Figure 1.**

3.3 Chemical structure of activated carbon

Activated carbon has a porous and crystalline structure. With this, it also has a chemical structure. The adsorption capacity of activated carbon is determined by its porous structure. But it is strongly influenced by a relatively small amount of chemically bonded heteroatom, mainly oxygen and hydrogen [15]. The variation in the arrangement of electron clouds in the carbon skeleton results in the creation of unpaired electrons and incompletely saturated valences which influence the adsorption properties of active carbons, mainly for polar compounds.

4. Synthesis of activated carbon

Up to date, commercial activated carbon (AC) used in wastewater treatment is produced from coals, woods, coconut shells, and lignite [16, 17]. Activated carbons possess several desirable properties that enable their use in adsorption. Properties, such as large surface area and porosity, together with surface chemistry react with molecules with specific functional groups. The wastewater treatment process is less profitable compared to other industrial sectors; it is always preferable to reduce the cost involved in its treatment process. The potential of low-cost adsorbent prepared from bio-waste has been identified in the last decade, and a great number of studies have been conducted to determine the characteristics and efficiencies of activated carbon produced from different bio-waste in the removal of different pollutants from wastewater. Synthesis of activated carbon from biomass generally starts with pretreatment of the sample, including crushing, drying at ~100°C, and sieving to obtain small particles within a specific size range. After these processes, the sample is carbonized in a dry inert atmosphere at 300–500°C, which promotes the elimination of volatile matters and tars and leads to the formation of biochar. Nowadays, the use of hydrothermal carbonization is attaining popularity in activated carbon production. In the hydrothermal process, the biomass is mixed along with water or reagent solution before carbonization [18].

The product obtained from hydrothermal carbonization is termed hydro char. Due to the different synthesis methods followed in the preparation of biochar, it is claimed that hydrothermal carbonization is more advantageous than traditional carbonization because the drying step carried out in the preliminary stage is not required. In such a process, a lower temperature of 180–250°C is used. The pressure released from the steam due to its closed system acts as an extra driving force to convert the biomass into hydro char. The formation of subcritical water under such conditions degrades cellulose, hemicelluloses, and lignin in the biomass [19]. The acidic gases, such as carbon dioxide (CO_2), nitrogen dioxide (NO_2), and sulfur dioxide (SO_2). eliminated during the heating will react with water to form an acidic solution. Therefore, the need to treat such gaseous pollutants is not required. The presence of several functional groups, especially oxygenated ones, on the hydro char was also found, which results in a higher adsorption capacity of contaminants [20]. The presence of functional groups improves the adsorption of heavy metals on the hydro char despite lower surface area compared to activated carbon [21]. After the carbonization of biomass, physical or chemical activations are required to activate the carbonized material. Physical activation is normally performed by passing inert gases, such as carbon dioxide (CO_2), nitrogen (N_2), or steam [22]. The gases are passed into the carbonized material at a high temperature of 700–900°C. Under these conditions, the conversion of carbonized material into CO_2 gas through oxidation is limited. Hence the yield of activated carbon is increased when compared to activation using air. Chemical activation proceeds only in the addition of activating reagents usually acid or base to the carbonous material. Former heating at 300–500°C is done, followed by washing the activated carbon to neutralize its pH. Potassium hydroxide (KOH) is one of the common basic reagents used in chemical activation. It inhibits tar formation in carbonized biomass [23]. In addition, KOH reacts with carbon in the precursor to form potassium carbonate (K_2CO_3), which then reacts further with carbon to form potassium (K), potassium oxide (K_2O), carbon monoxide (CO), and carbon dioxide (CO_2) [24]. These processes generate porosities in the adsorbents with large micropore volumes and narrower size distribution.

4.1 Preparation of activated carbon from agricultural waste biomass

Activated carbon is prepared from hemp stem [25]. The hemp stem was carbonized at 500°C in nitrogen (N_2) atmosphere for 1 hour. The carbonized material was then ground and mixed with potassium Hydroxide (KOH) solution for 24 hours, then it was dried and activated at 800°C in nitrogen (N_2) atmosphere. When zinc chloride ($ZnCl_2$) is added, it reacts with the char and governs the pore distribution during the heat treatment [26]. Acidic activating reagents are commonly used. Addition of phosphoric acid (H_3PO_4) to the carbonous material causes hydrolysis of glycosidic linkage in polysaccharides of hemicellulose and cellulose [27]. When phosphoric acid is used during chemical activation, it is possible to control the reaction of the acid with the carbonized biomass by utilizing the gases used. The resulting activated carbon prepared from hemp biomass will get an application of adsorbent for removing dyestuffs and also in wastewater treatment process.

Activated carbon prepared from spent tea leaves is mixed with phosphoric acid (H_3PO_4) and heated at 450°C in oxygen and air atmosphere [28]. Phosphoric acid

(H_3PO_4) reacts with oxygen and air atmosphere to form phosphorus oxides, then phosphorus oxides react with oxygen to form cerium oxide (CeO_2) electron pair bonds. Due to the extension of polyaromatic cross-linking, the higher porosity and high surface area could be observed in the formation of activated carbon. Phosphoric acid reacts in the air atmosphere to form phosphorus pentoxide (P_2O_5), and it sublimes from the activated carbon to increase its porosity development to a further extent. Activated carbon prepared by using steam activation in the atmosphere increases the deposition of carbon and decreases the porosity. When physical activation is compared to chemical activation, it requires a higher cost due to the chemical activating reagent and acid/base utilized for neutralizing pH in activated carbon. Consumption of energy will be lowered due to the lower temperature requirement during chemical activation [29]. Activated carbon prepared from chemical activation will have a higher surface area and well-developed porosity compared to the physical activation process [30]. Therefore, chemical activation is more widely followed in the preparation of activated carbon.

The merits of chemical activation over physical activation can be explained by the microstructure model [31]. This model states that every activated carbon material contains numerous micro domains in spherical shape, where the micropores develop. Mesopores, on the other hand, formed in inter micro domain space. Using physical (steam) and chemical activation (KOH), the phenol resin-based spherical carbon was converted to activated carbon via chemical activation. Activated carbon prepared by potassium hydroxide (chemical activation) exhibits a larger surface area, micropore volume, and average pore width. Whereas activated carbon prepared by physical activation by means of steam activation has less surface area, pore width, and micropore volume even at the same activation temperature. Surface area of the activated carbon material using a chemical process is 2878 m^2/g, at an activation temperature of 900°C, while activated by steam (physical activation) possesses a surface area of 2213 m^2/g. A lower yield of activated carbon occurs in chemical activation when a loss of carbon mass obtains during homogeneous pore development in the intra-micro domain regions. During physical activation using steam, homogeneous activation was observed, which results in lowering the efficiency of micropore development. Reduction of sizes of particles may also occur. Steam activation produces a lower yield of activated carbon with limited porosity, while chemical activation produces a uniform pore development.

Modifications are attempted in synthesizing activated carbon. The biomass is mixed with activating agents for chemical activation before pyrolysis. The pyrolysis process is carried out at high temperatures of 550–900°C. This method of preparing activated carbon is termed a one-step process. The preparation of activated carbon at a low temperature below 550°C after that activating agent is added for chemical activation. This process is termed as two-step pyrolysis process.

Activated carbon is prepared from corn stalk. One- and two-step processes of activation methods are followed for adsorbing cadmium. After the preparation and treatment of the contaminants, two-step pyrolysis processes increased the microporosity and surface area of the activated carbon. Therefore, the activated carbon prepared using two-step pyrolysis process shows higher adsorption capacity. There is no difference in its electrochemical properties. Activated carbon prepared from corn stalk shows its ability in removing cadmium from wastewater.

Activated carbon is prepared from *Cassia fistula* commonly called golden shower tree. A three-step process was done to synthesize activated carbon. The *C. fistula* was cleaned, dried, and crushed. The crushed sample was carbonized. The hydrochar obtained was further pyrolyzed to form biochar. Finally, it was activated by using potassium carbonate (K_2CO_3). After the application to adsorption process, the prepared activated carbon shows high performance in cationic dyes. It is evident that three-step preparation processes are more advantageous than using one-step and two-step processes. Morphological studies show that the adsorbent's character strongly depends on its preparation method. Activation process helps the adsorbent to increase its pore volume. Increase in pore volume enhances the adsorbent property to absorb more amounts of dyes. Preparation of Activated carbon using three-step processes reveals that it is a more effective method than other processes. Activated carbon prepared from *C. fistula* is a promising material for removing dyes from wastewater.

Activated carbon is prepared from *Tamarindus indica fruit shell*. It was washed with distilled water to remove its dirt and dried under sunlight to remove its water

S. No	Adsorbent	Activating agent	Temperature	Adsorption studies	Kinetic studies	Characterization
1.	Sugar beet pulp	H_3PO_4	Impregnated by 110°C for 12 hrs activated by (350, 400, 450, 500, and 550°C) time (0.5, 1, 1.5, 2, and 2.5 hrs)	—	Pseudo-second order	BET, FTIR
2.	Betel nut husk (BNH)	NaOH	BNH was mixed with powdered NaOH at three (1:1,1:2, and 1:3) ratios for 24 hrs at room temperature and activated by 500°C for 1 hr	Langmuir Freundlich	Pseudo-second order	SEM, FTIR
3.	Tamarind Shell (*Tamarindus indica fruit shell)*	NH_4Cl	Carbonized at 500°C for 2 hrs and activated for 24 hrs	Langmuir, Freundlich,	Pseudo-second order	SEM, FTIR,TGA/DTA
4.	Mango seed	$ZnCl_2$	Impregnated with 100°C for 30 mins, activated by 500°C for 1 hr.	Phenol adsorption	—	—
5.	Papaya seed	—	RPS washed with distilled water to remove the dust particles. It was left at room temperature for 1 day. Dried in oven at 105°C for 24 hrs	Langmuir	Pseudo-second order	—

Table 2.
Preparation of activated carbon from various agricultural waste biomasses, activating agent and its studies [32].

content. It is chemically activated by using ammonium chloride (NH_4Cl). The activated material is filtered and carbonized at 500°C for 2 hours [32]. The Langmuir model shows the formation of the adsorbent's monolayer coverage is at the adsorbent's outer space. While Freundlich model isotherm analysis confirms the monolayer adsorption capacity was high. The maximum dye removal percentage was 24.3 g l − [1]. From this, it is evident that the *T. indica fruit shell* biomass is a promising adsorbent for removing textile dyes and also it can be employed in wastewater treatment process. **Table 2** lists some of the biomass prepared from agricultural waste which can be employed in the water treatment process.

5. Applications of activated carbon

Activated carbons are proven to be effective in the removal of various pollutants from aqueous solutions, including dyes, pharmaceutical personal care products (PPCPs), heavy metals, and organic pollutants.

5.1 Removal of dyes from water resources

Dyes are one of the heavy pollutants which affect water bodies, due to the usage of dyes in paints, clothing, paper products, and plastics. In the textile industry alone, there are more than 3600 types of dyes used (Pure Earth and Green Cross Switzerland, 2017). Around 2–20% of dyes used for coloring in the textile industry are eliminating effluent. Therefore, the textile industry is the root cause of water pollution [33]. Due to the complex structure of dye molecules, the dyes do not degrade in water. Dyes mixed in water bodies, such as ponds, lake, and river, reduce the amount of sunlight reaching water sources, and due to that, photosynthesis gets affected for aquatic plants as well as animals [34]. Polluted water intake by humans also causes mutagenic and carcinogenic effects [35, 36]. Therefore, advanced studies are carried out on removing toxic dyes. For example, methylene blue (MB), a dye used in the textile industry causes complications in eyes, affects brain functions, and also causes skin diseases [37]. **Table 3** lists some of the biomass prepared from agricultural waste involved in removing dyes from water resources.

5.2 Removal of heavy metals from water

Heavy metals and anions present in drinking water become a challenging problem among the public due to their causes. The sources of these heavy metals and ions are paint industry, chemical plants, textile dying, dumping of waste in landfills, etc. Most

Source of biomass	Pollutants targeted to remove	Removal efficiency (%)
Bamboo cane powder (H_3PO_4/600°C, $H_2O + N_2$)	Lanasyn orange	2600
Cashew nutshell ($ZnCl_2$/400°C, N_2)	Methylene blue	476
Acorn shell ($ZnCl_2$/700°C)	Methylene blue	330
Lemon citrus peel (H_3PO_4/500°C)	Rhodamine B	254

Table 3.
Dyes adsorption on adsorbents derived from biomass for water treatment [36, 37].

Source of biomass	Pollutants targeted to remove	Removal efficiency (mg/g)
Loblolly pine chips (300°C, 93% N_2 + 7%O_2/NaOH/ 800°C, N_2)	Cd^{2+} (Cadmium)	167.3
Rice Husk (600°C/Na_2CO_3)	Pb^{2+} (Lead)	0.6
Banana peel (500°C, air/KOH/500°C, air)	Cu^{2+} (Copper)	13.24
Australian pinecones (800°C /NaOH)	Cu^{2+} (Copper)	12.82

Table 4.
Activated carbon from agricultural biomass in active removal of heavy metals from water [38, 39].

of the ions liberate from industries are toxic and carcinogenic to the environment as well as to humans. Consumption of water, which contains metal ions and anions, usually causes chronic effects instead of acute effects. Long-term disease will be affected by humans due to the usage of this contaminated water. Up to date the effects of heavy metals present in drinking water are identified in human body [38]. Despite the efficiency of adsorption in the removal of contaminants from wastewater, the use of commercially activated carbon is undesirable due to the low affinity towards heavy metals [39]. Therefore, it is vital to develop other adsorbents, especially from biomass waste, to minimize the heavy metal ions from entering the water bodies. **Table 4** lists out some of the biomass prepared from agricultural waste involved in removing heavy metal ions.

5.3 Removal of organic pollutants from water bodies

Palm oil is one of the main ingredients in cooking [40]. Usage of these oils globally, the economic growth raised in some countries like Malaysia and Indonesia, Around 39% of palm oil production is from Malaysia. Despite these huge benefits and economic credits, the removal of oil effluent is a major challenge. Palm oil mill effluent (POME) is a byproduct obtained after processing palm seeds. This effluent contains a high chemical oxygen demand (COD) and biological oxygen demand (BOD). Elimination of palm oil effluent in water sources affects aquatic lives due to the formation of harmful compounds in water. Palm oil mill effluent appears in black or brownish-colored slurry with a foul smell. The traditional method of treating Palm oil mill effluent was dumped in a large pit to degrade. It needs a large land area and long time to degrade [41]. This conventional method of treatment is not effective so the effluent will remain toxic. Therefore, adsorption is one of the best suitable techniques that can be applicable in removing oil effluent. **Table 5** represents some of the adsorbents prepared from agricultural waste biomass that effectively remove toxic compounds from oil effluents.

5.4 Pharmaceutical and personal care products (PPCPs) - pollutant removal

Due to the increase in diseases all over the world people are practicing or in- taking various kinds of drugs. The commonly used drugs are analgesics, antibiotics, anti-inflammatories, as well as painkillers [42]. Due to the large demand for drugs, production is getting increased, and also pharmaceutical waste is getting increased.

Source of biomass	Pollutants targeted to remove	Removal efficiency (%)
Sugarcane bagasse (700°C/KOH/600°C)	NH_3-N (Ammoniacal Nitrogen)	94.74% (color)
Sugarcane bagasse (700°C, N_2/ KOH/ microwave, N_2)	NH_3-N	97.83% (color)
Cow dung ash (CH_3COOH)	NH_3-N	79%
Bio sorbent from oil palm mesocarp fiber (600°C / steam/600°C)	palm oil mill effluent (POME)	80%

Table 5.
Activated carbon from agricultural biomass in active removal of organic pollutants from water bodies [40, 41].

Pharmaceutical and personal care products (PPCPs) are also the major cause of pollution, due to the disposal of waste in water sources. The toxic compounds liberated in water sources are consumed by wild animals and also human beings and cause various health issues. Biological activity of humans gets affected due to the consumption of drugs even in low concentrations [43]. Carbamazepine, naproxen, diclofenac, and ibuprofen are some of the drugs that are commonly practiced and cause biological effects in humans [44]. The contamination sources of pharmaceutical and personal care products (PPCPs) include hospital effluents, medical waste from factories during production of drugs, and waste due to consuming medicine discharged from the body and disposal of medicinal waste in landfills. **Table 6** represents the activated carbon synthesized from biomass to remove the medicinal waste in water.

5.5 Other applications of activated carbon

5.5.1 Removing contaminants in drinking water that add color, odor, and flavor

The surface of activated carbon helps in effective removal or adsorption of organic compounds. Change in the surface morphology of activated carbon removes flavor, odor, and color from drinking water. The activated carbon is calcinated at high

Biomass source	Target pollutant	Adsorption capacity mg/g
Waste tea residue (H_3PO_4/450°C/ steam)	Oxytetracycline	273.7
Sawdust ($ZnCl_2$/microwave)	Bisphenol A	334.28
Peach stone (H_3PO_4/400°C, air)	Carbamazepine (CBZ) Ciprofloxacin (CPX)	170.3 (Activated carbon-rice husk, CBZ) 113.0 (Activated carbon –Peach stone, CBZ)
Palm kernel shell (900°C, N_2 + CO_2)	Acenolol	0.69 mmol/g
Peach stone (600°C, air/300°C, air)	p-nitrophenol	234.3

Table 6.
Activated carbon from agricultural biomass in active removal of pollutants in pharmaceutical and personal care products (PPCPs) [43, 44].

temperatures to increase its surface area. After the carbon preparation, it is activated by steam. Due to this process, the surface area of the activated carbon increases. Adsorption property gets improved after the activation so that the odor and taste of the drinking water will be easily changed. The odor produced by the organic compounds and undesirable substance gets into the pores of the activated carbon. Therefore, the unwanted materials can be removed easily [45].

5.5.2 Decaffeination of coffee

Decaffeination is done to remove the extra content of caffeine present in coffee. Activated carbon is employed in removal of caffeine from coffee. Activated carbon is mixed with a solution of ethyl cellulose to adsorb ethyl cellulose. After this process, the activated carbon-containing ethyl cellulose is allowed to dry. Then, the aqueous coffee extract is added to the activated carbon-containing ethyl cellulose. This mixture will extract the caffeine present in the coffee. Activated carbon employed in decaffeination can be reused in the same application. This improves the cost-effective removal method and also cheaper adsorbent [45].

5.5.3 Refining sugar, honey, and candies

Refining or bleaching of cane sugar and honey is done in food industries. The bleaching process is made in liquid with a certain temperature to reduce its viscosity. Chemically activated carbons prepared from softwoods are best suited for treating darker syrups. The pH of the activated carbon is adjusted to neutral one to do this bleaching process. This refining method is cheaper with high efficiency [45].

5.5.4 Discoloration of liquors, juices, and vinegars

Activated carbon prepared from powdered wood is commonly used. The synthesis of activated carbon from powdered wood increases the size of the pores. An increase in porosity adsorbs the color molecules faster. Powdered wood-activated carbon can be directly added to liquors, juices, and vinegars. On constant stirring, the color will be adsorbed to the activated carbon. The contact time will depend upon the color of the compound. This discoloration method is effective cheaper and easy to process [45].

5.5.5 Water treatment in industries

Activated carbon is commonly used in water treatment industries. By using activated carbon-heavy metals, organic contaminants can be easily removed. Chemical activation will increase the surface and porosity of the activated carbon. Due to this, the adsorption process will occur faster and more effectively. Activated carbon from waste materials like agricultural biomass is used nowadays. This made the treatment process an economic one [45].

5.5.6 Tertiary wastewater treatment

Activated carbon with the mineral origin is more suitable for the treatment of wastewater. Due to the range of pore formation in activated carbon during the activation process, the adsorption behavior of the material increases. The contaminants present in the wastewater will be completely eliminated during the treatment process. The odor and color will also be removed [45].

5.5.7 Purification of air and industrial gases

Toxic gases emitted from industries can be treated by using activated carbon. It adsorbs the toxic material and organic pollutants present in the air. Coconut shell-activated carbon is commonly used because it is a micropore material. Greater granulometry than the coconut shell activated carbon is employed in water treatment process to avoid pressure drop. Chemically activated carbons are used to adsorb organic compounds. Standard activated carbon cannot retain organic compounds and acid gases like aldehydes, ammonia, or mercury vapors [45].

5.5.8 Compressed air purification (diving tanks and hospitals)

To fill oxygen or compressed air in a tank pump is essential. But the pumping equipment will release oils and impurities during pumping. The adsorption filters made using activated carbon are fixed in tanks and pumps to ensure that the air technically remains oil free. This effectively reduces petroleum-derived vapors. The coalescing filters and adsorption filters fixed in the pump will provide compressed air with the highest quality [45].

5.5.9 Recovery of gold, silver, and other precious metals

An activated carbon with a micropore can retain gold, silver, and precious metals. The right size of the adsorbent can give adsorption kinetics in accordance with hydraulics of the process which works better than other processes. The activated carbon should have a certain hardness to withstand acid elution, washing processes at various temperatures, and thermal reactivation [45].

6. Conclusion

Getting clean drinking water and also consuming becomes a major challenge nowadays. Therefore, recycling wastewater effectively fulfills these major problems. The most common problems faced in water treatment process are adsorbents. Adsorbents used in water treatment process are costly and also not available. Due to this, the entire process becomes expensive and non-profitable. Activated carbon prepared from agricultural biowaste can be used as an adsorbent. It has a high potential to replace commercial activated carbon in wastewater treatment processes due to its low cost and high performance. Advancement in adsorption process using biomass is a solution to the challenges of water scarcity. It will certainly lead to large-scale applications of activated carbons from renewable sources in wastewater treatment industry.

Acknowledgements

Dr. M. Jaya Rajan (Associate Professor, Department of Chemistry & Research) and Anish C.I (Research Scholar, Register No: 19113012031017) acknowledge the research center Annai Velankanni College, Tholayavattam-629157, Affiliated to Manonmaniam Sundaranar University, Tirunelveli- 627012, Tamil Nadu, India. providing support for this chapter.

Author details

Muthaian Jaya Rajan[1*] and Clastin Indira Anish[2]

1 Department of Chemistry and Research, Annai Velankanni College, Tholayavattam, Tamil Nadu, India

2 Department of Chemistry, Annai Velankanni College, Tholayavattam Affiliated to Manonmaniam Sundaranar University, Tirunelveli, Tamil Nadu, India

*Address all correspondence to: jayarajanm1983@gmail.com

References

[1] Bansal RC, Donnet JB, Stoeckli F. Active Carbon. New York: Dekker; 1988. p. 482

[2] Marsh H, Menendez R. Introduction to Carbon Science. 1st ed. United States: Butterworth-Heinemann; 1989. pp. 1-36. ISBN 9780408038379. DOI: 10.1016/B978-0-408-03837-9.50006-3

[3] Gonzalez MT, Molina-Sabio M, Rodriguez-Reinoso F. Steam activation of olive stone chars, development of porosity. Carbon. 1994;**32**:1407-1413

[4] Molina-Sabio M, Gonzalez MT, Rodríguez-Reinoso F, Sepulveda-Escribano A. Effect of steam and carbon dioxide activation in the micropore size distribution of activated carbon. Carbon. 1996;**34**:500-509

[5] Gonzalez MT, Rodriguez-Reinoso F, Garcia AN, Marcilla A. CO_2 activation of olive stones carbonized under different experimental conditions. Carbon. 1997;**35**:159-165

[6] Yang T, Lua AC. Characteristics of activated carbons prepared from pistachio-nut shells by physical activation. Journal of Colloid Interface Science. 2003;**267**:408-417

[7] Guo Y, Yang S, Yu K, Zhao J, Wang Z, Xu H. The preparation and mechanism studies of Rice husk based porous carbon. Materials Chemistry and Physics. 2002;**74**:320-323

[8] Lillo-Rodenas MA, Cazorla-Amoros D, Linares-Solano A. Understanding chemical reactions between carbons and NaOH and KOH an insight into the chemical activation mechanism. Carbon. 2003;**41**:267-275

[9] Tseng RL, Tseng SK, Wu FC. Preparation of high surface area carbons from corncob with KOH etching plus CO_2 gasification for the adsorption of dyes and phenols from water. Colloids and Surfaces. 2006;**279**:69-78

[10] Molina-Sabio M, Rodríguez-Reinoso F, Caturla F, Selles MJ. Development of porosity in combined phosphoric acid-carbon dioxide activation. Carbon. 1996;**34**:457-462

[11] Hu Z, Srinivasan MP. Mesoporous high-surface-area activated carbon. Microporous and Mesoporous Materials. 2001;**43**:267-275

[12] Hu Z, Guo H, Srinivasan MP, Yaming N. A simple method for developing Mesoporosity in activated carbon. Separation and Purification Technology. 2003;**31**:47-52

[13] Lay JJ, Fan KS, Chang JI, Ku CH. Influence of chemical nature of organic wastes on their conversion to hydrogen by heat-shock digested sludge. International Journal of Hydrogen Energy. 2005;**28**:1361-1367

[14] Zhang ML, Fan YT, Xing Y, Pan CM, Zhang GS, Lay JJ. Enhanced bio hydrogen production from cornstalk wastes with acidification pre treatment by mixed anaerobic cultures: Biomass. Bioenergy. 2007;**31**:250-254

[15] Chou CH, Wang CW, Huang CC, Lay JJ. Pilot study of the influence of stirring and pH on anaerobes converting high-solid organic wastes to hydrogen. International Journal of Hydrogen Energy. 2008;**33**:1550-1558

[16] Demirbas A. Agricultural based activated carbons for the removal of dyes from aqueous solutions: A review. Journal of Hazardous Materials. 2009;**167**:1-9

[17] Li W, Yang KB, Peng JH, Zhang LB, Guo SH, Xia HY. Effects of carbonization temperatures on characteristics of porosity in coconut shell chars and activated carbons derived from carbonized coconut shell chars. Industrial Crops and Products. 2008;**28**(2):190-198

[18] Islam MA, Sabar S, Benhouria A, Khanday WA, Asif M, Hameed BH. Nanoporous activated carbon prepared from karanj (Pongamiapinnata) fruit hulls for methylene blue adsorption. Journal of the Taiwan Institute of Chemical Engineers. 2017;**74**:96-104

[19] Kalderis D, Kotti MS, Mendez A, Gasco G. Characterization of hydrochars produced by hydrothermal carbonization of rice husk. Solid Earth. 2014;**5**(1):477-483

[20] Liu W, Wang X, Zhang M. Preparation of highly mesoporous wood-derived activated carbon fiber and the mechanism of its porosity development. Holz Forschung. 2017;**71**(5):363-371

[21] Zhou N, Chen H, Xi J, Yao D, Zhou Z, Tian Y, et al. Biochars with excellent Pb(II) adsorption property produced from fresh and dehydrated banana peels via hydrothermal carbonization. Bio resource Technology. 2017;**232**:204-210

[22] Selvaraju G, Abu Bakar NK. Production of a new industrially viable green activated carbon from Artocarpus integer fruit processing waste and evaluation of its chemical, morphological and adsorption properties. Journal of Cleaner Production. 2017;**141**:989-999

[23] Azmi NB, Bashir MJ, Sethupathi S. Anaerobic stabilized landfill leachate treatment using chemically activated sugarcane bagasse activated carbon kinetic and equilibrium study. Desalination and Water Treatment. 2016;**57**(9):3916-3927

[24] Basta AH, Fierro V, El-Saied H, Celzard A. 2-steps KOH activation of rice straw an efficient method for preparing high-performance activated carbons. Bio resource Technology. 2009;**100**(17):3941-3947

[25] Zhang B, Ma Z, Yang F, Liu Y, Guo M. Adsorption properties of ion recognition rice straw lignin on PdCl: Equilibrium, kinetics and mechanism. Colloids and Surface. 2017;**514**:260-268

[26] Caturla F, Molinasabio M, Rodriguezreinoso F. Preparation of activated carbon by chemical activation with $Zncl_2$. Carbon. 1991;**29**(7):999-1007

[27] Ahmed TF, Sushil M, Krishna M. Impact of dye industrial effluent on physicochemical characteristics of Kshipra River Ujjain City India. International Research Journal of Environmental Sciences. 2012;**1**(2):41-45

[28] Kan Y, Yue Q, Li D, Wu Y, Gao B. Preparation and characterization of activated carbons from waste tea by H_3PO_4 activation in different atmospheres for oxytetracycline removal. Journal of the Taiwan Institute of Chemical Engineers. 2017;**71**:494-500

[29] Tang CF, Shu Y, Zhang RQ, Li X, Song JF, Li B, et al. Comparison of the removal and adsorption mechanisms of cadmium and lead from aqueous solution by activated carbons prepared from Typhaangustifolia and Salix matsudana. The Royal Society of Chemistry Advances. 2017;7(26):16092-16103

[30] Zubrik A, Matik M, Hredzak S, Lovas M, Dankova Z, Kovacova M, et al. Preparation of chemically activated carbon from waste biomass by single

stage and two-stage pyrolysis. Journal of Cleaner Production. 2017;**143**:643-653

[31] Kim DW, Kil HS, Nakabayashi K, Yoon SH, Miyawaki J. Structural elucidation of physical and chemical activation mechanisms based on the micro domain structure model. Carbon. 2017;**114**:98-105

[32] Abisha BR, Anish C, Beautlin Nisha R, Daniel Sam N, Jaya Rajan M. Adsorption and equilibrium studies of methyl orange on tamarind shell activated carbon and their characterization. Phosphorus, Sulfur, and Silicon and the Related Elements. 2022;**197**(3):225-230

[33] Chequer FMD, de Oliveira GAR, Ferraz ERA, Cardoso JC, Zanoni MVB, de Oliveira DP. Textile dyes: Dyeing process and environmental impact. In: Eco-friendly Textile Dyeing and Finishing. London, United Kingdom: IntechOpen; 2013 [Online]. Available from: https://www.intechopen.com/chapters/41411 DOI: 10.5772/53659

[34] Regti A, Laamari MR, Stiriba S-E, El Haddad M. Potential use of activated carbon derived from Persea species under alkaline conditions for removing cationic dye from wastewaters. Journal of the Association of Arab Universities for Basic and Applied Sciences. 2017;**24**:10-18

[35] Alizadeh N, Mahjoub M. Removal of crystal violet dye from aqueous solution using surfactant modified $NiFe_2O_4$ as nano adsorbent isotherms, thermodynamics and kinetics studies. Journal of Nano Analysis. 2017;**4**(1):8-19

[36] Kalita S, Pathak M, Devi G, Sarma HP, Bhattacharyya KG, Sarma A, et al. Utilization of Euryale ferox Salisbury seed shell for removal of basic fuchsin dye from water equilibrium and kinetics investigation. The Royal Society of Chemistry Advances. 2017;7(44):27248-27259

[37] Ardekani PS, Karimi H, Ghaedi M, Asfaram A, Purkait MK. Ultrasonic assisted removal of methylene blue on ultrasonically synthesized zinc hydroxide nanoparticles on activated carbon prepared from wood of cherry tree experimental design methodology and artificial neural network. Journal of Molecular Liquids. 2017;**229**:114-124

[38] Jaishankar M, Tseten T, Anbalagan N, Mathew BB, Beeregowda KN. Toxicity, mechanism and health effects of some heavy metals. Interdisciplinary Toxicology. 2014;7(2):60-72

[39] Sigdel A, Jung W, Min B, Lee M, Choi U, Timmes T, et al. Concurrent removal of cadmium and benzene from aqueous solution by powdered activated carbon impregnated alginate beads. Catena. 2017;**148**:101-107

[40] Igwe J, Onyegbado C. A review of palm oil mill effluent (POME) water treatment. Global Journal of Environmental Research. 2007;**1**(2):54-62

[41] Kaman SPD, Tan IAW, Lim LLP. Palm oil mill effluent treatment using coconut shell based activated carbon adsorption equilibrium and isotherm. In: MATEC Web of Conferences. Vol. 2017. 2017. p. 03009. MATEC Web of Conferences 87, 03009 (2017) ENCON 2016

[42] Sun JT, Zhang ZP, Ji J, Dou ML, Wang F. Removal of Cr^{6+} from wastewater via adsorption with high-specific-surface-area nitrogen-doped hierarchical porous carbon derived from silkworm cocoon. Applied Surface Science. 2017;**405**:372-379

[43] Ebele AJ. Pharmaceuticals and personal care products (PPCPs) in

the freshwater aquatic environment. Emerging Contaminants. 2017;**3**(1):1-16

[44] Archer E, Petrie B, Kasprzyk-Hordern B, Wolfaardt GM. The fate of pharmaceuticals and personal care products (PPCPs) endocrine disrupting contaminants (EDCs) metabolites and illicit drugs in a waste water treatment works and environmental waters. Chemosphere 2017;**174**:437-446

[45] Carbotecnia [Internet]. 2021. Available from: https://www.carbotecnia.info/learning-center/activated-carbon-theory/activated-carbon-applications/?lang=en. [Accessed: December 19, 2021]

Section 2

Water Quality for Aquacultures

Chapter 4

Water Quality and Aquatic Ecosystem Assessment Using Water Quality Indices in West Africa: Challenge and Perspectives

Lallébila Tampo, Idrissa Kaboré,
Seyf-Laye Alfa-Sika Mande and Limam Moctar Bawa

Abstract

"Water quality is health" as it is said, "water is life." The water quality and aquatic ecosystem health assessment is essential for early detection of water habitat degradation and overall aquatic ecosystem disturbances. All water management programs/agencies need simple and cheaper tools for water quality assessment. However, in West Africa there is an urgent need of water quality assessment tools, as far they are very limited. This chapter presents water quality indices as simple and cost-effective tool to monitor water quality. We explore the categories of water quality indices (WQIs), their use/application, and their scope in West African countries. We found that some multimetric indices (MMIs) are developed in West Africa, but they are not well known/used by local water managers and decision makers. There are not yet biotic indices (BIs) and physicochemical water quality indices (PCQWIs) typical to Western African countries areas, but most of them are applied/adapted to meet the needs of West African water quality and ecosystem health assessment. In this chapter, we present the results of some studies led in some West African countries regarding water quality indices (MMIs, BIs, PCWQIs), and address challenges and perspectives for long-term management of water and biological resources in developing countries.

Keywords: water quality, indices, West Africa, water management, aquatic ecosystem, pollution

1. Introduction

It is often said "water is life" but it must also be said, "water quality is health." Water is one of the most important resources for life to sustain our planet earth. It requires proper attention in terms of quantity as well as quality. Until the late 1960, the interest in water has been the available amount for consumption, except when poor water quality conditions persist, the available water was considered acceptable for consumption. Only during this last century, water quality has been deteriorated by human pressures to the point where this is now considered as a big concern for all nations [1]. In our modern society, the population and industrial growth with the establishment of the populations nearby water catchments and sources are exacerbating the deterioration of water quality [2]. The increasing demand of water resources has also contributed to the change in ecosystems functioning directly through human activities and indirectly by the non-point source pollution. The quality of water is threatened by a large number of pathogens [3], as well as anthropogenic chemical release from municipalities that enter into the water cycle [4]. In addition, discharges from municipal and industrial waste treatment plants have been identified as major sources of aquatic ecosystem pollution in industrialized countries [5]. Other pollution sources of water, including agricultural activities (e.g., using chemical fertilizers and pesticides), atmospheric deposition, industrial and mining wastes/activities, marine dumping, may heavily affect water resources and health of all aquatic ecosystems. Therefore, various pollutants may end up into the water column and degrade its quality [6]. Another problem that we observe recently in coastal areas such as southern Togo (coastal sedimentary basin) is the increase of salinity of groundwater due to seawater intrusion in coastal aquifers [7]. In the context of fast population growth and increasing water sources pollution, the measurements/assessment of water quality and aquatic ecosystem health is essential for early detection of water quality and habitat degradation, and may help to reinforce the preservation of water and biological resources. Furthermore, water quality measurement is essential for the comparison of data worldwide and can help to solve issues in decision making in terms of water resources management policies. Even if in the developed countries, the quality management tools are developed, and there is a strong need for education and training of water managers and users. These gaps have been recognized by European Commission that fund a series of training courses covering several topics such as monitoring and measurements of lake ecological status, heavy metals and organic compounds in waters, and the use of biological indices. These approaches contribute to supply good surface water and groundwater over the Europe for many uses. In general, despite improvements of water management laws in recent decades, access to good quality water remains a critical issue, and water quality is paramount for public health. Water and aquatic ecosystems pollution is a societal concern around the globe [6]. According to Nobel Laureate Richard Smalley (1996 Nobel in Chemistry), good water is the second challenge next to energy, among the humanity's top 10 problems in the next 50 years. Of all the global struggles around environmental protection and restoration, the quality of water and aquatic ecosystem health may be the most significant challenge and opportunity in the anthropocene. Water quantity and quality (access and management) are interlinked to global biohealth for the maintaining the well-being of a sustainable environment for plants, animals, and humans. Therefore, information about water quality is crucial for guiding efforts, and to reduce waterborne illnesses, identifying risks of disturbances and improving sanitation programs [8]. However in developing countries such as West Africa, many constraints have been identified in water monitoring and management including poor regulatory enforcement and insufficient resources, equipment and logistical, as well as the weakness of traditional analysis based on physicochemical, which is also very expensive. However, the detection of water quality and establishment of monitoring programs require not only creation of water agencies but also simple and low-cost evaluation tools. Also, the urgent need of local water agencies to manage and establish a restoration plan for polluted water and impaired aquatic ecosystems remains a challenge. This chapter aims at giving a background in terms of water quality measurements such as water quality indices (WQIs) in West Africa. Therefore, the chapter presents the adaptation or development of WQIs (methodology, study area/region/country and context of use) in some countries/regions of West Africa, and we raised challenges and perspectives for sustainable water resource management in West Africa.

2. Concept of water quality

Water quality is commonly and globally defined by its physical, chemical, and biological characteristics. Water may be used for drinking, irrigation, industrial processes, production of fish, shellfish and crustaceans, as well as habitats for wildlife including animals and plants [9]. The water quality depends on its uses. Indeed, the quality may be good enough for drinking but not suitable for other use. It may be good for irrigation of some crops but not suitable for irrigation of other crops. It may be well for livestock but not suitable for fish culture, whereas water quantity is determined by a single parameter (e.g., water mass or water volume), water quality depends on its component, as well as everything the water might have picked up during its runoff [1]. It is closely linked to the surrounding environment and land use. The modification of natural stream flows and the temperature can also have a major impact on water quality. Groundwater is a major source of water and should be away from contamination sources (e.g., urban or industrial, wastes dumping). The waters including streams/rivers and lakes waters are habitats for many aquatic ecosystem organisms. An ecosystem is a community of organisms including plants, animals, fungi, and bacteria, interacting with one another, and with their environment [9]. Protecting aquatic ecosystems is therefore important to maintain water integrity. The aquatic components are an integral part of our environment and play an important role in maintaining water quality, and are often use as valuable indicator of water quality. The physical characteristics (temperature, turbidity, Secchi disk depth, color, salinity, suspended solids, dissolved solids, etc.), chemical characteristics (dissolved oxygen, biological oxygen demand, chemical oxygen demand, pH, nutrients, heavy metals, hardness, alkalinity, etc.), and biological characteristics (algae/total chlorophyll, total biomass, macrophytes, bacteria, macroinvertebrates, fish, etc.) of water are worldwide used in the bioassessment, and were considered as water quality indicators.

Thus, there are many approaches to assess/describe the quality of given water (e.g., using its physical, chemical, or biological characteristics). One way to describe the quality of a water sample is to list out the concentrations of everything that the sample contains following the water quality standard guideline. The second way is a simplification of water quality data by aggregating the measurements of water quality parameters/indicators in a single number (water quality index (WQI)) to expresses overall water quality. Between the two ways, the WQIs are more beneficial and adapted for water quality control. Therefore, the formulation and use of indices have been strongly advocated by agencies [1]. Furthermore, once the WQIs are developed and applied, they serve as a convenient tool to examine trends, and to highlight specific environmental conditions [10, 11], and to help governmental decision makers in monitoring the effectiveness of regulatory programs.

3. Concepts of indices and use of water quality indices (WQIs)

The indices are representations of a condition/situation derived from a combination of several relevant parameters/measurements. The combination leads to a single ordinal number (e.g., score) that helps to understand and interpret water status [1]. The concept of using an index to represent a single value is not a novel idea. For example, these approaches have been used in others domains, such as economics and commerce [12]. Indices have also been used in ecology to represent species richness, evenness, diversity, etc. Also, in numerous others fields such as medicine, sociology, process safety, indices are extensively used. The environmental indices have also been used in life cycle assessment [13] and to characterize different types of environmental damages, including global warming. However, the major component of environmental indices including water quality indices (WQIs) is used as communication tool by regulatory agencies to describe the "quality" or "health" of water/aquatic ecosystems [10, 11], as well as to assess the impact of regulatory policies on various environmental management practices [14, 15]. The water quality indices provide a simple method for expressing the quality of water or the health of aquatic ecosystem. The significance of the WQIs can be appreciated as environmental performance indicators or holistic environmental performance index [16]. Water quality indices can be constructed in two ways: (i) increasing scale indices in which the index scores/numbers increase with the degree of pollution and (ii) decreasing scale indices in which the index scores/numbers decrease with the degree of pollution. The first one can be considered as "water pollution indices" and the second one as "water quality indices." However, the difference between the two ways/types can be seen as essentially cosmetic. Indeed, "water quality" is a general term of which "water pollution" indicates "undesirable water quality," is a specific case. Based on water quality parameters/indicators that are incorporated in an index to judge water quality, the WQIs can be vaguely classified into two categories: i) indices predominantly based on physical and chemical characteristics/features and ii) indices based on bioassessment or biological characteristics/features. In this chapter, the first category of indices will be named physicochemical water quality indices (PCWQIs). They incorporate one or more microbiological parameters (e.g., fecal coliforms, total coliforms, *Escherichia coli*) and predominantly based on physicochemical parameters/indicators. The second category of WQIs named biological water quality indices (BWQIs) is based on sampling, identification, enumeration of biological organisms including macroinvertebrates, fish, algae, macrophytes, etc.

3.1 WQIs based on bioassessment/biological water quality indices (BWQIs)

The assessment of water quality and aquatic ecosystem health using biological approaches is based on investigation of biota. However, it is necessary to identify indicator organisms that reflect water quality or aquatic ecosystem health, but not only due to the vagaries of chance. These organisms must be easily observed and counted. Therefore, the organisms or group of organisms used can include fish, amphibians, bacteria, protozoans, diatoms, algae, macrophytes, macroinvertebrates, etc. Only BWQIs based on macroinvertebrate features are discussed in this chapter. There are three types of BWQIs as follows:

i. *Biotic indices (BIs)*: are indices of water pollution/quality based on a study of the biota. Biotic indices often refer to the scoring biotic indices. But the term biotic index is vague, very wide, and may include diversity indices and comparison (similarity or dissimilarity) indices.

ii. *Indices of biotic integrity* or *indices of biological integrity* (IBIs), also called *multimetric indices*: are indices, which incorporate suits indices or metrics rather than a single index.

iii. *Multivariate approaches for bioassessment of water quality:* are methods, which use statistical tools to develop relationships between fauna and environmental characteristics for an "ideal" or high-quality reference site.

3.1.1 Biotic indices (BIs)

3.1.1.1 Background

The Saprobien system of Kolkwitz and Marsson used in Germany rivers in the early 1900s has been generally considered as the first biological scoring system so the first biotic indices for water quality assessment [17, 18]. Indices based on the Saprobien System are determined by the presence and absence of specific species that are sensitive to organic pollution. Thus, this concept used in Germany was expanded to other countries making it the first-water quality index. The development of that index is welcome since human activities produced an unprecedented pollution harmful to water and the biota [19, 20]. In 1964, the Trent Biotic Index (TBI) and other several modern biotic indices were developed in USA [21]. In the subsequent years, we have seen a slowly increasing reliance on biotic indices as a water quality management tool, especially in the developed countries. The biological indices are increasingly becoming a key element of environmental management policies and water resource in most developed countries [22, 23]. However, the use of biotic indices is not yet well known in developing countries. In many African countries, particularly in West African countries, there is no standardized or accredited biotic index used at regional scale. In recent years, most modern biotic indices are based on benthic macroinvertebrates [24]. To provide cost-effective tool accessible locally for ecological assessments, there is an increasing emphasis on the use of BIs [25, 26]. BIs based on macroinvertebrates are developed or applied in many countries including some developing countries such as South Africa, whereas in West Africa, there is a deficiency of information on the use of macroinvertebrates and biotic indices for water quality and ecosystem health assessment [27–33].

3.1.1.2 Approaches to BIs formulation

Like the physicochemical quality-based indices, the biotic indices can be used to communicate in an understandable way to natural resource managers, decision makers, politicians, and the general public [34]. The biotic indices are "response based" approaches for environmental monitoring wherein the strategy is to assess the overall aquatic environmental health. It involves the monitoring of biological or ecological indicators to characterize the response of the environmental disturbances. The disturbance of an aquatic ecosystem or a water body can be monitored *via* some factors that alter the biotic integrity, such as chemical variables, physical features, hydrology, energy source [35, 36]. The scope of response can result from the environmental modification. Then, the value of the response of each taxon is estimated as a tolerant/sensitive (score value), and each particular group of taxon is assigned to a sensitivity weighting or a score to particular pollutants or pollution. The score generally varies from 0 to 10. Like all WQIs, the score of a taxon increases with the increase of the sensitivity. To determine the value of biotic index in sampling site, the scores of all the individual taxa sampled at this site are summed and/or averaged to provide a valuable value of ecological health of the community; thus, the health of the water body, can be gauged. Even if the scoring system varies, most of modern and accredited biotic indices were adapted from [37] formula (1):

$$BI = \frac{\sum n_i s_i}{N} \quad (1)$$

where n_i is the number of specimens in each taxonomic group, s_i is the score for that taxonomic group, and N is the total number of organisms in the sample.

3.1.2 Indices of biological integrity (IBIs)

3.1.2.1 Background

The first IBI was introduced by [38] and was based on fish before being extended to other organisms. The IBI used only attributes of fish assemblage to assess the condition of rivers and its catchment. This index of biotic integrity reflects water-land linkages (e.g., water quality and land use), physical habitat quality, hydrological regime, and biological interactions [39, 40]. The IBI was designed to integrate information from individual, population, assemblage, and ecosystem levels into a single numerical indicator. Initially proposed by [38] and later improved by [41, 42], the IBI combines 12 fish assemblage attributes/metrics in three categories: (i) species richness and composition, (ii) trophic composition, and (iii) fish status and abundance. These data are used to assess sites condition by comparing reference sites "undisturbed" and "disturbed sites" within the same ecoregion [43] or in a similar ecoregion [44]. When there are no appropriate undisturbed sites, the least-disturbed regional sites can be considered [44] or the reference condition can be modeled from knowledge of historical data and fish habitat preferences [44]. This index was first developed for streams in the Midwestern of United States and has been proven useful in many other regions of North America [45]. It has also been applied to estuaries [46], lakes [47], and rivers outside the United States and Canada [48]. However, even if an IBI follows an ecoregional approach, it was often adapted to fulfill the requirement at regional scale. But these adapted IBIs are not always successful. For example, a modified IBI in semiarid southeastern Colorado, USA, was not able to reflect habitat degradation [49]. Nevertheless, over the years, a number of shortcomings of Karr's IBI have been identified and attempts have been made to overcome those infirmities. In the process, new IBIs have been developed based not only on fish, but also on macroinvertebrates and diatoms. Interestingly, despite the variety of adaptations throughout the world, the fundamentals of the original IBI still stand strong. In West Africa, some IBIs based on macroinvertebrates are developed for water quality and ecosystem health assessment during these last years [10, 11, 50–52].

3.1.2.2 Fundamentals and principles to (IBIs) formulation

The IBIs, also called indices of biotic integrity, incorporate a suitable metrics in a single index called multimetric index (MMI). IBIs use a combination of indices in an endeavor to assess multiple anthropogenic pressures on aquatic ecosystems. The undisturbed ecosystems support an unbalanced biological condition over time. The organisms that inhabit a natural system, both individually and community level, can be used as potential indicators of ecosystem conditions because their presence as well as their well-being can be influenced by the human-induced perturbation [53].

Whereas biotic indices seek the representation of a natural water body through certain species or group of species, IBIs metrics are chosen to reflect the taxonomic composition, trophic relationships, abundance, and condition of organisms within an aquatic community [54, 55]. Thus, IBIs aim to convey a more integrated picture of ecosystem health than BIs. Due to the multiple metric combinations in IBIs, we can suppose that IBIs are more sophisticated than BIs. IBIs can reflect important components of ecology: taxonomic richness, habitat, and trophic guild composition, besides individual health and abundance. Differences in expected species richness and composition associated with different regions or basins, water body sizes and location in drainage are factored into metric selection and scoring. The main focus of the IBIs is based on the assemblage structure (indicator organisms or combinations of organisms) such as fish, plankton, benthos, and macrophytes rather than ecosystem processes, and yet both structural and functional metrics are often included in the IBIs [56]. IBIs provide a perception of ecological assemblage integrity to common people. Assessment of biological integrity using the IBIs requires "reference habitat" to compare the successful of index to discriminate human perturbation.

3.1.2.3 Approaches and steps to the development of IBI

Building a robust and effective IBI is based on proper selection of measurable attributes that provide reliable and relevant signals about the effects of human pressures on biota. The biological attributes (metrics) ultimately incorporated into an IBI should reflect specific predictable responses of organisms to landscape condition modification. These metrics have to be sensitive to a range of physical, chemical, and biological features. They should be relatively easy to measure and interpret. A typical IBI include several attributes of biota, including taxa richness, indicator groups, health of individual organisms, and ecological processes. The most important criterion for choosing a biological attribute as a metric is whether the attribute responds predictably along a gradient of human influence. An effective IBI comprising well-chosen metrics should integrate information from ecosystem, community, and individual levels and clearly discriminate the biological "signal" including the effects of human activities from natural variation. The main steps to the development of IBI are described as follows:

Selection of candidate metrics.

This selection is based on the consideration of ecological relevance as well as feasibility of measurement. Candidate metrics should include measures of species diversity, productivity (abundance and biomass), and tolerance to anthropogenic stress. Barbour et al. [57] recommended to group macroinvertebrate metrics by categories, such as taxonomic richness, taxonomic composition, tolerance/intolerance, feeding group (e.g., predators, scrapers and filter feeders) and habit type (e.g., clingers and burrowers), as well as life history. When the ecological relevance is based on very specific ecological concepts, candidate metrics are basically a priori selected. However, indices developed with a utilitarian approach typically begin with a large list of candidate metrics, which is then pruned.

Selection of core and relevant metrics.

This step consists in retaining among ecologically relevant candidate metrics, those that are more sensitive (responsive to anthropogenic action, both degradative and restorative), and more representative (able to measure status and trends relative to policy decisions and management actions). The core metric nomination/selection is based on community level and characters that represent key community aspects.

This selection protocols include the following: (i) *a priori* selection based on a specific ecological foundation and/or best professional judgment; (ii) selection based on univariate statistical tests comparing undergraded and degraded samples from calibration data [58]; and (iii) utilitarian selection based on multivariate tests using a calibration dataset [59, 60]. Otherwise, many steps can be considered for relevant metrics selection.

Metric range: The first hurdle, which a candidate metric must overcome, is the range test. The "range" is the distribution of metric values across all available data, and the goal of the range test is to identify metrics that provide a large range of dataset. Metrics that have very small ranges (e.g., richness metrics based on only few taxa) or the ones that have similar values (e.g., most sites have values = 0) should be excluded.

Reproducibility: A metric providing fairly reproducible values at individual sites is more useful than a metric, which is less precise due to its variation within sites. Low sampling variation is necessary if a metric is to have a high probability of discriminating good and poor sites [61].

Adjusting for natural gradients: Metric values can vary with both stressor and natural gradients (e.g., elevation, slope, and stream size). Thus, knowledge of how to allocate the variability in metric values between natural and anthropogenic gradients is important. Selection of metrics that seem to respond strongly to stressor but, in fact, are merely correlated with the same natural gradient should be avoided [61]. One of the techniques to normalize metrics for natural gradients is to remove the stressor gradient from the data by focusing solely on reference sites data and to quantify the remaining correspondence between the metric value and the natural gradient.

Responsiveness: The efficiency of a metric is directly linked with its ability to distinguish degraded from relatively undisturbed sites. This can be tested in many ways. For example, metrics can be chosen on the basis of their correlation with specific stressors (e.g., nutrients, organic pollution, human pressures, and sedimentation). Some of the original metrics used by Karr [38] were chosen on the basis of their theory responses to specific aquatic stressors. However, several difficulties occur when metrics are evaluated in terms of relationships with specific stressors. First, many stressors are strongly correlated with one another, and attributing metric response to any particular stressor could inflate the role of this stressor. Second, not all stressors are well quantified (e.g., short-lived pesticides or herbicides), or even known, at all sites. In the absence of pristine sites that may serve as frames of reference, the evaluation of the responsiveness of the metric is based on the metric ability to distinguish least-disturbed (reference) from the most-disturbed sites. Metric scoring and typical threshold selection are based on a set of least-disturbed sites.

Final metric selection and check for metric redundancy: All metrics that are successful with discriminatory power may be included in the IBIs. Candidate metric that is the most discriminating is chosen first and then preceded iteratively by adding the most responsive metric from each metric category until all categories are represented. This iterative process is based on the assumption that choosing the most responsive individual metrics will provide the most robust IBI such as each metric provides unique information (i.e., that the metrics are not redundant).

Metric redundancy can be defined at least in two different ways: (i) Metrics provide very similar biological information or (ii) metrics are highly correlated with other metrics. The first of these definitions may be important, because one would be disinclined to include two metrics that are based on identical (or broadly overlapping) biological or taxonomic information [61]. However, apparent redundant metrics might be, in fact, relatively uncorrelated. Thus, one might avoid including metrics with values that are strongly correlated. If two metrics co-vary because the same taxa are changing in abundance as levels of disturbance rise and fall, then the metrics are effectively correlated. However, if metrics co-vary because they respond to similar stressors, then the correlated metrics are not necessarily redundant.

Metric combination.

The most difficult challenge in index development is not only the selection of metrics, but also the combination to capture the dynamics of essential ecological processes, and metric easy use by water managers. Without a strong and obvious ecological foundation, an index will not be relevant and therefore difficult to be used. The first step in metrics combination is to normalize selected metrics. In this chapter, we present two methods for metrics scoring: (i) discrete scoring (with values such as 1, 3, or 5, based on a subjective assessment of the range of each metric) and (ii) continuous scoring (with calculated values using some formula). Discrete scoring can mask subtle differences since and forces the scores to be in one or the other discrete interval. This could dampen the ability of the index to differentiate ecological condition classes [62, 63]. The continuous scoring can avoid the subjective nature of discrete scoring. The use of continuous scoring requires two considerations: how to decide what values of a metric indicate "ideal" biological condition (e.g., a scores 1, 10, or 100) and what values indicate unacceptably bad condition (e.g., a score of 0). The 95th percentile of the reference site distribution of values for each metric is generally used as the scoring ceiling and the 5th percentile of the distribution of values at all sites as the scoring floor. This approach produces an IBI with the highest responsiveness and lowest variability. Metric values between the ceiling and floor are interpolated linearly to yield intermediate values, and the final MMI for a site is calculated as the sum of its scored metrics. When interpreting final scores, the IBI is generally rescaled to range, for example, from 0 to 100, 0 to 10, or 0 to 1. The final MMIs, which are generally the sum or the mean of normalized values of selected metrics, follow or adapt the general formula below:

$$MMI = \sum Selected\ metric\ scores \tag{2}$$

MMI validation.

The robustness of any IBI depends on the effectiveness of metric (e.g., widely and repeatedly). Index validation should ideally include the following: (i) testing of the index with different from the data set used in index development; (ii) setting a correct *a priori* classification criteria, and (iii) *a posteriori* criterion based on best professional judgment. More commonly after index development, new data are used as validation data. Independent validation, by scientists other than those proposing the index, is highly appreciated. Some degree of intercalibration or validation can also be achieved by determining the level of agreement provided by an index with the best professional judgment [64] or by comparing the level of agreement between indices of different geographical origin, such as when comparing results of indices from Europe and USA.

3.1.3 Multivariate approach for bioassessment

The multivariate approaches for bioassessment are methods that use statistical tools to develop relationships between biota and environmental characteristics of "ideal" or reference site. Thus, the relationships are used to predict the pattern of fauna distribution. The observed fauna at the test site is then compared with predicted fauna. The usual elements of multivariate analyses often include the cluster analyses, ordination techniques, and discriminant analyses [1, 65]. In contrast, IBIs follow inductive approach that relies on a posteriori assumption. A number of multivariate approaches have been standardized and consequently adopted for widespread use at international level. For example, the first of widely used multivariate approach is the River Invertebrate Prediction And Classification System (RIVPACS) developed in the UK. The RIVPACS approach has been adopted in many countries up to now [65, 66]. However, no multivariate approach is developed and even applied/adapted in West African countries. Therefore, this approach is not discussed in this chapter.

3.2 WQI based predominantly on physicochemical characteristics (PCWQIs)

The water quality indices based predominantly on physicochemical characteristics are often called in the most of literature "water quality indices" (WQIs). This leads to confusion with other indices such as BWQIs. In order to avoid this confusion, in this chapter, WQIs based predominantly on physicochemical characteristics will called "physicochemical water quality indices"(PCWQIs).

3.2.1 Background

PCQWIs using a numerical scale began with Horton's index in 1965. Horton used three main criteria to select variables/parameters for its combination in index: (i) The number of variables to be handled by the index should be limited to avoid making the index unwieldy; (ii) the variables should be of significance in most areas; and (iii) only variables of which reliable data are available, or obtainable, should be included. Based on these criteria, Horton selected 10 most commonly measured water quality variables for index formulation. Horton's index did not include any toxic chemicals and was highly subjective as they are based on the judgment of the author and associates [67]. Thus, several authors have built less and less subjective PCWQIs based on pioneering works. The useful PCWQIs incorporate up to 14 variables including toxic chemicals, physical parameters, microbiological parameters, etc. Nowadays, there is increasing use of ordination (e.g., factor analysis, principal component analysis), and other concepts such as entropy and genetic algorithms in making "hybrid" indices or in enhancing the applicability of PCWQIs.

3.2.2 Steps and approaches of PCWQIs formulation

Four steps are often used in the development of a PCWQI: (i) parameters selection; (ii) transformation of the parameters of different units and dimensions to a common scale; (iii) assignment of weightages to all parameters; and (iv) aggregation of subindices to produce a final index score. However, additional steps may also be taken to improve the index efficiency.

3.2.2.1 Parameter selection

A PCWQI would become unwieldy if each and every possible physical and chemical parameter is included in the index. As possible constituents of water are paramount, parameter selection is as fraught with uncertainty and subjectivity, and this is crucial for the usefulness of any index. Enormous efforts, attention, experiences, and consensus-gathering skills are needed to ensure the most relevant parameters included in a PCWQI. Therefore, studies have suggested to involve large number of experts opinion, WHO standards/criteria, and statistical approaches in order to attempt the reduction of the subjectivity in the index building.

3.2.2.2 Transformation of the parameters of different units and dimensions to a common scale: Making subindices

Most of water are expressed in different units and, therefore, have different behavior in terms of concentration-impact relationship. Thus, before index formulation, all parameter units have to be transformed into a single scale varying usually from 0 to 1. Some index scales might have the range of 0–100. This step is a standardization or normalization of water quality parameters using functions called subindices.

Subindices development.

As described in [1], for each selected parameter to be integrated in the final index their units and range of concentrations (from highly acceptable to highly unacceptable) are transformed in a single scale. For example, if one considers a set of n pollutant variables denoted as (x_1, x_2, x_3, x_i, x_n), then for each pollutant variable x_i, a subindex I_i is computed using subindex function $f_i(x_i)$. To develop the final index, based on a sound available data, some expert consensus and mathematical formula, authors suggest to use the subindex function as follows:

$$I_i - F_i x_i \tag{3}$$

Once the subindices are defined/calculated, they usually are aggregated together in a second mathematical step to form the final index as follow:

$$I = g(I_1 + I_2 + \dots I_n) \tag{4}$$

The aggregation function Eq. (4) usually consists of either a summation operation, in which individual subindices are added together or aggregated to form the final index as illustrated in **Figure 1** following [1].

Different types of subindices.

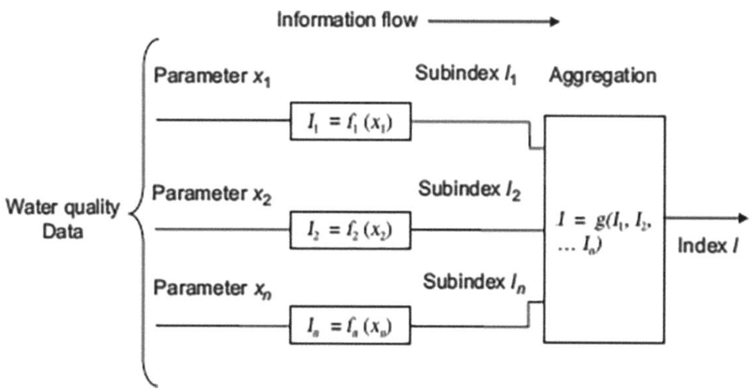

Figure 1.
The index development process [1].

Subindices can be classified as one of four general types: linear; nonlinear; segmented linear; segmented nonlinear. In this chapter, only the linear subindices are considered/applied following the model of Eq. (5):

$$I = \alpha x + \beta \tag{5}$$

where I is the subindex, x the pollutant variable, and α and β the constants. With this function, a direct proportion exists between the subindex and the pollutant variable. The linear indices are simple to compute and easy to understand.

3.2.2.3 Assignment of weightages

Even if all short-listed parameters are deemed to be important as water quality indicators, but they do not have the same weights. Within the selected parameters, some would have a high importance than others. There are some indices, which assume equal weightage for all the parameters. However, in many cases, different weightage is given to the different parameters. The assignment of weightage is, like selection of parameters, therefore, subjective and so, need well-formulated techniques of opinion and expert's consensus.

3.2.2.4 Aggregation of subindices to produce a final index

Several methods are often used for final index formulation. For example, the two most basic ones are:

Additive: The subindices are combined through summation (e.g., arithmetic mean). This is the commonly used method.

Multiplicative: In multiplicative aggregation, the subindices are combined through product operation (e.g., geometric mean).

In this chapter, only additive aggregation is addressed including linear sum index or a weighted linear sum index.

Linear sum index.

A linear sum index is often computed by the addition of unweighted subindices following the model of Eq. (6):

$$I = \sum_{i=1}^{n} I_i \tag{6}$$

where I_i is the subindex for pollutant variable i and n is the number of pollutant variables.

Weighted linear sum index.

A weighted linear sum index can be given by Eq. (7):

$$I = \sum_{i=1}^{n} W_i I_i \tag{7}$$

where I_i is the subindex for i^{th} variable and W_i the weight for i^{th} variable with:

$$\sum_{i=1}^{n} W_i = 1 \tag{8}$$

4. Illustrative examples of BWQIs in West African countries

4.1 Multimetric indices (MMIs)

In this section, some illustrative examples of MMI developed in West Africa are featured. They cover different biogeographic regions in terms of climate and water bodies' types. The summaries of these illustrative examples are focused on the methodologies, and brief discussion of their validation and application.

4.1.1 A multimetric index of Zio river basin in Togo (MMIZB)

4.1.1.1 Methodology

This study focuses on Zio River basin, one of the basins with many permanent flow streams and drained by several streams coming mainly from Togo Mountains. The basin is characterized by a tropical climate including the main ecoregions of Togo (moist forests, gallery forests and a mosaic of agricultural land, degraded forest, and savanna).

The study followed human pressures gradient ranging from least impaired sites (references) to impaired sites. The impaired sites are characterized by intense agriculture and other human activities described in [68]. Forty-two sites were sampled from downstream to upstream according to the following criteria: (i) minimally impaired sites (reference sites); (ii) fairly or slightly impaired sites (intermediate sites); and (iii) severely impaired sites. Environmental variables were used to compare differences between reference and impaired sites using Mann-Whitney comparison test, and seasonal aspects on environmental variables were assessed using the Kruskal-Wallis comparison test ($p < 0.05$). The analysis procedures including candidate metrics examination, core metrics selection, and final multimetric index formulation is detailed in [10].

Lumb et al. [10] have demonstrated that high correlation coefficients ($r > 0.80$, $p < 0.05$) were interpreted as indicating redundant metrics and in these cases, only one metric (the most usual metric in tropical climate and strongly correlated with environmental variables) was retained for multimetric index development.

To fulfill all requirements described previously, the continuous scoring was used to score/standardize the metric values. The scoring procedure of MMIZB index was achieved in four steps with two formula following Eqs. (9) and (10): (i) computing all six core metrics; (ii) scoring the metrics Eq. (9) for metrics that decrease with increasing impairment, and Eq. (10) for metrics that increase with increasing impairment; (iii) applying a simple interpolation to adjust values to a range from 0 to 100; (iv) final multimetric index value is obtained by summing scores of selected metrics, and classes boundaries were set to promote local use of index by managers.

$$Standard\ metric = \frac{Metric\ result - 25th\ percentile\ of\ impaired\ sites}{75th\ percentile\ of\ reference\ sites - 25th\ percentile\ of\ impaired\ sites} \times 10 \tag{9}$$

$$Standard\ metric = \frac{Metric\ result - 75th\ percentile\ of\ impaired\ sites}{25th\ percentile\ of\ reference\ sites - 75th\ percentile\ of\ impaired\ sites} \times 10 \tag{10}$$

4.1.1.2 MMIZB index validation and application

The validation test of the MMIZB index sensitivity was conducted with 11 new sites of Zio River basin following steps described in [10]. These sites were chosen following some reference conditions approach described in [57] with local experts' consensus and were not used in the building process of MMIZB index. The test was performed using a principal component analysis (PCA) of 13 key environmental variables. The assessment of the index sensitivity was done using pressures scores of first axis of the PCA, and organic pollution of Zio river basin was assessed using Prati's Index [69]. The multimetric macroinvertebrates index of Vietnam (MMI_Vietnam) [70] was also used to determine the robustness of MMIZB index. This sensitivity of the index was also tested by assessing whether there was a clear discrimination among reference sites and impaired sites using whisker box plots.

MMIZB was developed in a watershed/basin (the main hydrological unit) covering three of the five ecoregions of Togo. Furthermore, it was validated by the data from the three ecoregions, MMI of Vietnam (MMI of a tropical basin), and water quality index of organic pollution. MMIZB is then considered as one of the first relevant multimetric index developed at a watershed level in the tropical regions of West Africa. Therefore, MMIZB may be used/applied for watersheds/basins ecosystem health and water quality assessment in Togo, West Africa, and other tropical countries.

4.1.2 The Sahel River multimetric index (SRMI) of West African Sahel Rivers, Burkina Faso

4.1.2.1 Methodology

Burkina Faso is a sub-Saharan, landlocked country in the central part of West Africa. Most of Burkina Faso lies within the West Sudanian Savannah. The climate is tropical and semi-arid and characterized by a north-south gradient in rainfall distribution, with high variability in time and space. The study was undertaken in the three main basins in Burkina Faso: Nakanbé (formerly White Volta) in the central part of the country, Mouhoun (formerly Black Volta) in the west, and Comoé in the south-west part of Burkina Faso. The Volta river is composed of Nakanbé and Mouhoun at country scale. The Mouhoun river has permanent run-off, and the Nakanbé downstream has two hydropower dams, which have almost perennial flows see [11]. According to Refs. [11, 29], the catchment of these rivers is under pressure from various human activities that lower the ecological integrity of rivers.

Sampling sites selected within the three basins covered a gradient from protected areas "slightly impaired" to highly impaired sites [11, 36]. Six reference sites were defined as "natural or near-natural sites" following a priori criteria: no disturbance or near-to-natural hydro-morphological features, preserved natural habitats, no human activity within 100 m of the riparian zone, dissolved oxygen (DO) > 6 mg/L, conductivity <75 μs/cm, no sand or gravel excavation, no visible sign of sensory features, natural vegetation cover typical to area > 80%, and the presence of wild birds and mammals possibly. The selected impaired sites were identified as those in agricultural or urbanized riparian zones with collapsed and eroded riverbanks. The impaired sites were also exposed to point and non-point sources of pollution. For environmental variables measured at reference and impaired sites, seasonal and temporal variability was assessed using the Mann-Whitney comparison test [11].

The authors followed the same procedures: candidate metrics examination, core metrics selection as described in Ref. [10], except the final multimetric index formulation. The selected core metrics were normalized using Eq. (11) for metrics that decrease with impairment and Eq. (12) for metrics that increase with impairment. The continuous scoring was used to avoid distortion of scores by potentially extreme maximum values and sample size noise. Therefore, 95th and 5th percentile thresholds were applied. The scores of each normalized core metric were scaled from 0 to 1. Finally, the individual metrics were summed by aggregating the scores of each normalized core metric to form the SRMI. A value close to 1 represents high ecological status, and a value close to 0 represents bad ecological status. The practical relevance for water management was considered when the numerical range of the SRMI (0–1) was grouped into five ecological quality classes: (I) high ecological quality class, (II) good ecological quality class, (III) fair ecological quality class, (IV) poor ecological quality, and (V) bad ecological quality class based on literature and experts consensus.

$$\text{Normalised value } (Vi') = \frac{V_i}{V_{95\%}} \tag{11}$$

$$\text{Normalised value } (Vi') = \frac{V_{max} - V_i}{V_{max} - V_{5\%}} \tag{12}$$

where Vi' is the normalized value of the metric, V_i is the metric value, $V_{95\%}$ is the 95th percentile, $V_{5\%}$ is the 5th percentile, $V_{\max}$ is the maximum value of the metric.

4.1.2.2 SRMI validation and application

Kaboré et al. [11] have proven sensitivity of their index with data covering most of ecoregions of West Africa. The same authors have shown that the SRMI responded to a set of environmental parameters associated with a gradient of human pressures affecting the ecological integrity of water bodies ($R^2 \geq |0.50|$; $p < 0.05$; $p < 0.001$). This confirms the usefulness of an unprecedented and promising tool for biological monitoring and decision making in Sahelian regions' water management [11]. Therefore, SRMI may be used/applied for water quality and aquatic ecosystem health assessment in other sahelian countries or regions of Africa.

4.1.3 Others MMIs of West Africa countries

In North Central Nigeria, Edegbene et al. [50] have formulated a macroinvertebrates multimetric index named "River Chanchaga multimetric index" (MMIchanchaga) to assess the ecological status of River Chanchaga. The authors used several number of core metrics (13) for final index formulation compared to those of [10, 11]. At the same time, Edegbene et al. [51] have produced macroinvertebrates tool for urban river systems assessment in Delta region of Nigeria using five core metrics for the final index formulation. Additionally, a macroinvertebrate-based multimetric index has been also formulated by [52] for assessing ecological condition of forested stream sites draining Nigerian urbanizing landscapes. The mutimetric is a relevant sophisticated tool for ecosystems health assessement in West Africa, because it is acknowledged advantage of combining the sensitivity of many metrics to different aspects. However, Kaboré et al. [11] argued that develop a less sophisticated method; for example, a biotic score remains a challenge for West African limnologists.

In regard of all these advantages proved by macroinvertebrate, we encourage West Africa limnologists to use macroinvertebrates at family taxonomy resolution for bioassessment and biomonitoring program implementation due to their cost-benefit (e.g., ease of sampling and identification of specimens) and limited taxonomic knowledge of the local water managers.

4.2 Promising tools using macroinvertebrates for ecosystem health assessment in West Africa

Freshwater biodiversity is threatened by climate and land use change. Freshwater ecosystems depend strongly on physical and chemical features such as water quantity, water quality, water flow, and surrounding vegetation. Freshwater ecosystems throughout the world are threatened by human activities that directly alter hydrology system, such as construction of physical barriers (e.g., for dams' and against floods), water extraction, and filling or draining of shallow habitats. Pollution of waterways with toxic substances and excessive nutrients, as well as destructive land use practices in surrounding freshwater ecosystems, leads to alterations of water quality. In West Africa, the impacts on water bodies are expected to increase due to high levels of economic and population growth in this region. Effects of those anthropic multiples pressures on water ecosystems are rendering benthic macroinvertebrate assemblages among others aquatic organisms more vulnerable including change of community composition, increase of opportunistic species number, and decline of sensitive taxa and general biodiversity. In West Africa, several authors such as Tampo et al. [28], Kaboré et al. [29], Kaboré et al. [30], Tampo et al. [31], Agblonon Houelome et al. [32], and Edegbene et al. [33] among others have demonstrated the sensitivity of macroinvertebrates to water quality and human disturbances. From these authors, the macroinvertebrates community including functional surrogates, taxonomic composition, and diversity reflect the environmental condition, and any modification from physical, chemical, and biological integrity of freshwater ecosystems can dramatically affect water and biological resources, and as human health.

These results revealed that macroinvertebrate assemblages have high potential use as ecological indicators making them particularly beneficial for bioassessment because: i) they are the most popular indicators and their use dates back to the late 1840; ii) they are the major group of organisms in terms of species richness and individual abundance in most water bodies; iii) their life cycles are sufficiently long that they will likely be exposed to pollution and environmental stress; iv) sampling the benthic macroinvertebrates assemblage is relatively simple and does not require complicated devices or great effort; v) although they are mobile, they have mostly sedentary habits so they are likely to be exposed to local pollution or environmental stress; vi) the benthic macroinvertebrates biology are well known, and thus, sufficient identification keys, ecological data bases, and methodological standards are existent.

5. Examples of PCWQIs application in West African countries

As mentioned for BIs, there is not yet a standard PCWQI developed for West Africa countries. However, water quality is assessed in West Africa countries by using/adapting PCWQIs developed at abroad far from African reality/context. We present here some examples of studies applying/adapting PCWQIs in West Africa.

5.1 Assessment of water quality using PCWQIs in Nigeria

Many studies reported the application/adaptation of PCWQIs for the assessment of water quality in several water sources across Nigeria [71, 72]. The application/adaptation of PCWQIs commonly based on the following: (i) the most common measured water quality variables (temperature, electrical conductivity, biochemical oxygen demand, fecal coliform, pH, dissolved oxygen, total phosphates, turbidity, nitrates and total solids, total hardness, etc.); (ii) water quality rating scale; (iii) relative weight, and (iv) overall PCWQI calculation. Some PCWQIs are often calculated by averaging the individual index values of some or all of the parameters within five water quality indicator categories (water clarity, dissolved oxygen, oxygen demand, nutrients, and bacteria) that depicts the pollution level or status of the water. Other PCQWI commonly applied/adapted in Nigeria is based on Canadian Council of Ministers of Environment (CCME). These methods combined three factors into a single index [73]. Several authors have applied theses PCWQIs to evaluate the water quality from different water sources especially surface and groundwater across the different zones of Nigeria. According to Adelagun et al. [71], PCWQIs can be applied to evaluate the water quality from different water sources across different zones. Etim et al. [74] have carried out a study to compute a PCWQI in order to assess the suitability of water from different sources collected from different areas in the Niger Delta region of Nigeria. From Ref. [74], the PCWQI has been calculated using standard drinking water quality recommended by the World Health Organization (WHO) and Indian Council for Medical Research (ICMR).

5.2 Assessment of water quality using PCWQIs in Ghana

According to Ref. [75], using PCWQI as a tool for water quality assessment has become one of the new ways to disseminate scientific information to the general public and policy makers in Ghana. The concept of WQI to describe the state of water quality in Ghana has not been popular until 2003, when the water resource commission (WRC) produced an adapted WQI document and proposed the WQI concept for assessing surface water quality in Ghana [75]. Indeed, in 2003 the WRC produced a document (Ghana Raw Water Quality Guidelines and Criteria: Adapted Water Quality Index) based on Solway Water Quality Index, like an index to characterize the overall raw water quality in Ghana. The Adapted Water Quality Index is a classification system that uses an index from selected water quality parameters. The index classifies water quality into one of the four categories: good, fairly good, poor, and grossly polluted. Each category describes the state of water quality compared to objectives that usually represent the natural state. Thus, the index indicates the degree to which the natural water quality is affected by human activities. From Ref. [75], the concept of a water quality index for Ghana arose from two needs. The first is to share and communicate with the public, in a consistent way. The second is to provide a general means of comparing and ranking various water bodies throughout Ghana. In the same country, Miyittah et al. [76] have used the Canadian Council of Ministers of the Environment WQI (CCMEWQI) approach to assess the pollution status of Aby Lagoon System.

Banoeng-Yakubo [77] and Boateng et al. [78] have also conducted studies using PCWQI to analysis groundwater quality of the northern section of the Volta region. Authors defined steps for computing the PCWQI. For example, in the first step, each of the chemical parameters was assigned a weight (wi) based on their perceived effects on primary health. The highest weight of five was assigned to parameters, which have the major effects on water quality in the study area [78].

5.3 Assessment of water quality using PCWQIs in Togo

The assessment of water quality using PCWQIs is not well known or is almost at its beginning in Togo. However, water quality indices have been applied to assess water quality or suitability for drinking and irrigation purposes [7], and for the validation of other water quality indices [10]. We present here two examples of study cases applying PCWQIs. One study for water quality assessment in a peri-urban area conducted by [7] and another conducted by [79] on groundwater in the Southeastern Togo. Indeed, based on some monitoring data, Tampo et al. [7] conducted a study in a peri-urban area, where PCWQIs have been used to compare the suitability of three types of water (treated wastewater, and ground and surface water) for domestic and irrigation purposes. The authors adapted three types of PCWQIs such as Prati's index of pollution (PIP), national sanitation foundation's water quality index (NSFWQI), and overall index of pollution (OIP) for the comparison and classification of waters. This study used physicochemical and microbiological parameters as indicators to incorporate in the PCWQIs. Another study conducted by Napo et al. [79] in the Southeastern Togo adapted PCWQIs methods to assess the groundwater water quality. The PCWQI was calculated by aggregation subindices of 13 physicochemical parameters in linear sum index.

6. Water quality assessment: challenges and perspectives in West Africa

There are a few multimetric indices developed in West Africa, but they are not well known by local population, and are less used or not at all applied for water quality and aquatic ecosystem health assessment. In this context of exacerbating human pressures on freshwater ecosystems, there is an urgent needed to share these findings with water users and decision makers, because the effectiveness of any index-driven management policies will increase with better knowledge of conservation status. But, how to link resilience of climate change and environmental conservation with ensuring livelihoods of a growing population remains a great challenge. All the funds are not using efficiently and mostly directed to the primary needs (e.g. food, security, heath) leading to conceal the environment problems. Others hands, there is a need for an applied education program, on several scales "bottom up and top down" on environmental issues, as well as, for academic, water agencies, and NGOs to use scientific knowledge in theory and practice. Often, the lack of clear mainstreams cooperation between university and ministries on water quality assessment methods leads to waste investments and lower researches capacity in terms of development and policy basis to sustainable management of aquatic ecosystems in West Africa. Therefore, we encourage all researches that provide interdisciplinary national data incorporating social and political components on water quality assessment. This may significantly strengthen national capacities in the improvement of water management. For less sophisticated such as scoring biotic indices, out of perspectives argued by Kaboré et al. [11] there are no biotic indices developed for Western Africa region. The findings from authors such as Tampo [10], Kaboré et al. [11], Tampo et al. [31], Agblonon Houelome et al. [32], Edegbene et al. [50–52], among others, have confirmed the advantages of using macroinvertebrates features because they are effective tools and respond to a gradient of multiple human pressures affecting the ecological integrity of freshwater ecosystems. Other authors such as Tampo et al. [7], Adelagun et al. [71], Iwar et al. [72], Banoeng-Yakubo et al. [77], and Napo et al. [79] found that PCWQI score based on the main/important physicochemical parameters can be used to classify water from excellent and good or bad for consumption and crop irrigation. All these studies reveal the importance of PCWQIs application for water quality monitoring in West Africa, but some standards/guidelines about water quality are needed to be intensively explored and determined at each country level in order to refine PCWQIs application in West African countries.

With ongoing severe freshwater pollution and multiple human pressures, the need of tools accessible locally is crucial for prioritizing conservation efforts and efficient management of freshwater ecosystems in Western Africa region. Funds to develop methods for aquatic ecosystems health assessment in West Africa remain weak. Thus, more cooperations between NGOs, local government, stakeholders, and international partners are still need to reinforce the capacity building of developing countries for an integrative water management and governance when society meets ecology.

7. Conclusion

In West Africa, there still is a lack of both water management tools and water agency. The creation of water management agency may be encouraged by the development of cheaper tools for water quality assessment. Water quality indices seem to be an alternative in Africa where traditional analysis is not only expensive but also inaccessible. The water quality indices can be used as effective tools for ecological awareness in West Africa. There are an urgently call for new co-creation of water management tools including water managers and agency, decision makers, as well as water users. This may help to internalize findings and their use for long-term management of water resources, the overall biota, ecosystems functioning, and services for human well-being.

Author details

Lallébila Tampo[1,2*], Idrissa Kaboré[3], Seyf-Laye Alfa-Sika Mande[1,2,4] and Limam Moctar Bawa[2]

1 Faculty of Science and Technology, University of Kara, BP, Togo

2 Laboratory of Applied Hydrology and Environment, Faculty of Sciences, University of Lomé, BP, Togo

3 Laboratory of Animals Biology and Ecology (LBEA), 03 BP 7021, University Joseph KI-ZERBO, Ouagadougou, Burkina Faso

4 Beijing Key Laboratory of Water Resources and Environmental Engineering, China University of Geosciences (Beijing), Beijing, P.R. China

*Address all correspondence to: charlestampo@gmail.com

References

[1] Abbasi T, Abbasi SA. Water Quality Indices. Amsterdam: Elsevier; 2012. p. 362. DOI: 10.1016/B978-0-444-54304-2.00016-6

[2] Vorosmarty CJ, McIntyre PB, Gessner MO, Dudgeon D, Prusevich A, Green P, et al. Global threats to human water security and river biodiversity. Nature. 2010;**467**:555-561. DOI: 10.1038/nature09440

[3] Rizzo L, Manaia C, Merlin C, Schwartz T, Dagot C, Ploy MC, et al. Urban wastewater treatment plants as hotspots for antibiotic resistant bacteria and genes spread into the environment: A review. Science of the Total Environment. 2013;**447**:345-360. DOI: 10.1016/j.scitotenv.2013.01.032

[4] Schwarzenbach RP, Escher BI, Fenner K, Hofstetter TB, Johnson CA, von Gunten U, et al. The challenge of micropollutants in aquatic systems. Science. 2006;**313**(5790):1072-1077. DOI: 10.1126/science.1127291

[5] Reemtsma T, Weiss T, Mueller J, Petrovic M, Gonzalez S, Barcelo D, et al. Polar pollutants entry into the water cycle by municipal wastewater: A European perspective. Environmental Science & Technology. 2006;**40**(17): 5451-5458. DOI: 10.1021/es060908a

[6] Voudouris K, Voutsa D, editors. Water Quality Monitoring and Assessment. London, UK, Rijeka, Croatia: InTech; 2012. p. 602. DOI: 10.5772/2411

[7] Tampo L, Alfa-Sika Mande SL, Adekanmbi AO, Boguido G, Akpataku KV, Ayah M, et al. Treated wastewater suitability for reuse in comparison to groundwater and surface water in a peri-urban area: Implications for water quality management. Science of the Total Environment. 2022;**815**: 152780. DOI: 10.1016/j.scitotenv.2021.152780

[8] Peletz R, Kisiangani J, Bonham M, Ronoh P, Delaire C, Kumpel E, et al. Why do water quality monitoring programs succeed or fail? A qualitative comparative analysis of regulated testing systems in sub-Saharan Africa. International Journal of Hygiene and Environmental Health. 2018;**221**:907-920. DOI: 10.1016/j.ijheh.2018.05.010

[9] Postolache O, Girão PS, Pereira JMD. Water quality monitoring and associated distributed measurement systems: An overview. In: Voudouris K, Voutsa D, editors. Water Quality Monitoring and Assessment. Rijeka, Croatia: InTech; 2012. pp. 25-64. DOI: 10.5772/32159

[10] Tampo L, Lazar IM, Koboré I, Oueda A, Akpataku VK, Djaneye-Boundjou G, et al. A multimetric index for assessment of aquatic ecosystem health based on macroinvertebrates for the Zio river basin in Togo. Limnologica. 2020;**83**:125783. DOI: /10.1016/j.limno.2020.125783

[11] Kaboré I, Ouèda A, Moog O, Meulenbroek P, Tampo L, Bancé V, et al. A benthic invertebrates-based biotic index to assess the ecological status of West African Sahel Rivers, Burkina Faso. Journal of Environmental Management. 2022;**307**:114503. DOI: 10.1016/j.jenvman.2022.114503

[12] Diewert WE, Nakamura AO. Indices. In: Essays in Index Number Theory. Vol. 1. Amsterdam: North Holland Press; 1993. pp. 71-104

[13] Khan AA, Paterson R, Khan H. Modification and application of the

Canadian Council of Ministers of the Environment Water Quality Index (CCME WQI) for the communication of drinking water quality data in Newfoundland and Labrador. Water Quality Research Journal of Canada. 2004;**39**(3):285-293. DOI: 10.2166/WQRJ.2004.039

[14] Pusatli OT, Camur MZ, Yazicigil ZH. Susceptibility indexing method for irrigation water management planning: Applications to K. Menderes river basin Turkey. Journal of Environmental Management. 2009;**90**(1):341-347. DOI: 10.1016/j.jenvman.2007.10.002

[15] Sadiq R, Haji SA, Cool G, Rodriguez MJ. Using penalty functions to evaluate aggregation models for environmental indices. Journal of Environmental Management. 2010; **91**(3):706-716. DOI: 10.1016/j.jenvman.2009.09.034

[16] Lumb A, Sharma TC, Bibeault JF. A review of genesis and evolution of water quality index (WQI) and some future directions. Waters Quality, Exposure and Health. 2011;**3**:11-24. DOI: 10.1007/s12403-011-0040-0

[17] Washington HG. Diversity, biotic and similarity indices. A review with special relevance to aquatic ecosystems. Water Research. 1984;**18**:653-694. DOI: 10.1016/0043-1354(84)90164-7

[18] Rico E, Rallo A, Sevillano MA, Arretxe ML. Comparison of several biological indices based on river macroinvertebrate benthic community for assessment of running water quality. Annals Limnology. 1992;**28**:147-156. DOI: 10.1051/limn/1992013

[19] Davis WS, Simon TP, editors. Biological Assessment and Criteria: Tools for Water Resource Planning and Decision Making. London: Lewis Publishers; 1995. p. 432. DOI: 10.13140/RG.2.1.4916.2726

[20] Perry J, Vanderklein E. Water Quality: Management of a Natural Resource. USA: Wiley-Blackwell; 1996. p. 656. DOI: 10.2307/2405269

[21] Woodiwiss FS. The biological system of stream classification used by the Trent River board. Chemistry and Industry. 1964;**83**:443-447

[22] Norris RH, Norris KR. The need for biological assessment of water quality: Australian perspective. Australian Journal of Ecology. 1995;**20**:1-6. DOI: 10.1111/j.1442-9993.1995.tb00516.x

[23] Moog O, Chovanec A. Assessing the ecological integrity of rivers: Walking the line among ecological, political and administrative interests. Hydrobiologia. 2000;**422**:99-109. DOI: 10.1023/A:1017053829050

[24] Ollis DJ, Dalls HF, Esler K, Boucher C. Bioassessment of the ecological integrity of river ecosystems. African Journal of Aquatic Science. 2006;**31**:205-227. DOI: 10.2989/16085910609503892

[25] Brown CA. A comparison of several methods of assessing river condition using benthic macroinvertebrate assemblages. African Journal of Aquatic Science. 2001;**26**:135-147. DOI: 10.2989/16085910109503735

[26] Metzeling L, Chessman B, Hardwick R, Wong V. Rapid assessment of rivers using macroinvertebrates: The role of experience, and comparisons with quantitative methods. Hydrobiologia. 2003;**510**:39-52. DOI: 10.1023/B:HYDR.0000008500.34301.a0

[27] Arimoro FO, Muller WJ. Mayfly (Insecta: Ephemeroptera) community

structure as an indicator of the ecological status of a stream in the Niger Delta area of Nigeria. Environmental Monitoring and Assessment. 2010;**166**: 581-594. DOI: 10.1007/s10661-009-1025-3

[28] Tampo L, Oueda A, Nuto Y, Kabore I, Bawa L, Djaneye-boundjou G, et al. Using physicochemical variables and benthic macroinvertebrates for ecosystem health assessment of inland rivers of Togo. International Journal of Innovation and Applied Studies. 2015;**12**: 961-976

[29] Kaboré I, Moog O, Alp M, Guenda W, Koblinger T, Mano K, et al. Using macroinvertebrates for ecosystem health assessment in semi-arid streams of Burkina Faso. Hydrobiologia. 2016; **766**:57-74. DOI: 10.1007/s10750-015-2443-6

[30] Kaboré I, Ouédraogo I, Tampo L, Ouéda A, Moog O, Guenda W, et al. Composition and dynamic of benthic macroinvertebrates community in semiarid area rivers of Burkina Faso (West Africa). International Journal of Biological and Chemical Sciences. 2016; **10**:1542-1561. DOI: 10.4314/ijbcs.v10i4.8

[31] Tampo L, Kaboré I, Alhassan EH, Ouéda A, Bawa LM, Djaneye-Boundjou G. Benthic macroinvertebrates as ecological indicators: Their sensitivity to the water quality and human disturbances in a Tropical River. Frontiers in Water. 2021;**3**:662765. DOI: 10.3389/frwa.2021.662765

[32] Agblonon Houelome TM, Agbohessi TP, Adandedjan D, Nechifor R, Chikou A, Lazar IM, et al. Ecological quality of the Alibori River, northern Benin, using macroinvertebrate indicators. African Journal of Aquatic Science. 2022;**2**(47):178-184. DOI: 10.2989/16085914.2022.2044749

[33] Edegbene AO, Odume ON, Arimoro FO, Keke UN. Identifying and classifying macroinvertebrate indicator signature traits and ecological preferences along urban pollution gradient in the Niger Delta. Environmental Pollution. 2021;**281**: 117076. DOI: 10.1016/j.envpol.2021.117076

[34] Stark JD. SQMCI: A biotic index for freshwater macroinvertebrate coded-abundance data. New Zealand Journal of Marine and Freshwater Research. 1998; **32**:55-66. DOI: 10.1080/00288330.1998.9516805

[35] Abbasi T, Abbasi SA. Water quality indices base on bioassessmentethe biotic indices. Journal of Water and Health (IWA Publishing). 2011;**9**(2):330-348. DOI: 10.2166/wh.2011.133

[36] Kaboré I, Moog O, Ouéda A, Sendzimir J, Ouédraogo R, Guenda W, et al. Developing reference criteria for the ecological status of West African rivers. Environmental Monitoring and Assessment. 2018;**190**:2. DOI: 10.1007/s10661-017-6360-1

[37] Hilsenhoff WL. Use of arthropods to evaluate water quality of streams. Wisconsin Dept. of Natural Resources Technical Bulletin. 1977;**100**:15

[38] Karr JR. Assessment of biotic integrity using fish communities. Fisheries. 1981;**6**:21-27. DOI: 10.1577/1548-8446(1981)006<0021:AOBIUF>2.0.CO;2

[39] Steedman RJ. Modification and assessment of an index of biotic integrity to quantify stream quality in Southern Ontario. Canadian Journal of Fisheries and Aquatic Sciences. 1988;**45**:492-500. DOI: 10.1139/f88-059

[40] Allan JD, Erickson DL, Fay J. The influence of catchment land use on stream integrity across multiple spatial scales. Freshwater Biology. 1997;**37**: 149-161. DOI: 10.1046/j.1365-2427.1997.d01-546.x

[41] Fausch KD, Karr JR, Yant PR. Regional application of an index of biotic integrity based on stream fish communities. Transactions of the American Fisheries Society. 1984;**113**: 39-55. DOI: 10.1577/1548-8659(1984)113<39:RAOAIO>2.0.CO;2

[42] Karr JR, Fausch KD, Angermeier PL, Yant PR, Schlosser IJ. Assessing biological integrity in running waters. A method and its rationale. Illinois Natural History Survey Special Publication 5. 1986. p.28

[43] Omernik JM. Ecoregions of the conterminous United States. Annals of the Association of American Geographers. 1987;**77**:118-125. DOI: 10.1111/j.1467-8306.1987.tb00149.x

[44] Hughes RM. Defining acceptable biological status by comparing with reference conditions. In: Davis WS, Simon TP, editors. Biological Assessment and Criteria: Tools for Water Resource Planning and Decision Making. Boca Raton: Lewis Publishers; 1995. pp. 31-47

[45] Simon TP, Lyons J. Application of the index of biotic integrity to evaluate water resource integrity in freshwater ecosystems. In: Davis WS, Simon TP, editors. Biological Assessment and Criteria: Tools for Water Resource Planning and Decision. Boca Raton: Lewis Publishers; 1995. pp. 245-262

[46] Thompson BA, Fitzhugh GR. A Use Attainability Study: An Evaluation of Fish and Macroinvertebrate Assemblages of the Lower Calcasieu River, Louisiana. Baton Rouge, USA: Center for Wetland Resources. Coastal Fisheries Institute, Louisiana State University; 1986. p. 143

[47] Minns CK, Cairns VW, Randall RG, Moore JE. An index of biotic integrity (IBI) for fish assemblages in the littoral zone of Great Lakes areas of concern. Canadian Journal of Fisheries and Aquatic Sciences. 1994;**51**:1804-1822. DOI: 10.1139/f94-18

[48] Hughes RM, Oberdorff T. Applications of IBI concepts and metrics to waters outside the United States and Canada. In: Simon TP, editor. Assessment Approaches for Estimating Biological Integrity Using Fish Assemblages. Lewis Press: Boca Raton; 1998. pp. 79-83

[49] Bramblett RG, Fausch KD. Variable fish communities and the index of biotic integrity in a western Great Plains river. Transactions of the American Fisheries Society. 1991;**120**:752-769. DOI: 10.1577/1548-8659(1991)120<0752:VFCATI>2.3.CO;2

[50] Edegbene AO, Elakhame LA, Arimoro FO, Osimen EC, Odume ON. Development of macroinvertebrates multimetric index for ecological evaluation of a river in north Central Nigeria. Environmental Monitoring and Assessment. 2019;**191**:274. DOI: 10.1007/s10661-019-7438-8

[51] Edegbene AO, Arimoro FO, Odume ON. Developing and applying a macroinvertebrate based multimetric index for urban rivers in the Niger Delta, Nigeria. Ecology and Evolution. 2019;**9**: 12869-12885. DOI: 10.1002/ece3.5769

[52] Edegbene AO, Akamagwuna FC, Odume ON, Arimoro FO, Ovie TTE, Akumabor EC, et al. A macroinvertebrate-based multimetric index for assessing ecological condition

of forested stream sites draining Nigerian urbanizing landscapes. Sustainability. 2022;**14**:11289. DOI: 10.3390/su141811289

[53] Mack J. Developing a wetland IBI with statewide application after multiple testing iterations. Ecological Indicators. 2007;**7**:864-881. DOI: 10.1016/j.ecolind.2006.11.002

[54] Karr JR, Yoder CO. Biological assessment and criteria improve total maximum daily load decision making. Journal of Environmental Engineering. 2004;**130**:594-604. DOI: 10.1061/(ASCE)0733-9372(2004)130:6(594)

[55] Dolph CL, Sheshukov AY, Chizinski CJ, Vondracek B, Wilson B. The index of biological integrity and the bootstrap: Can random sampling error affect stream impairment decisions? Ecological Indicators. 2010;**10**:527-537. DOI: 10.1016/j.ecolind.2009.10.001

[56] Schindler DW. Experimental perturbations of whole lakes as tests of hypotheses concerning ecosystem structure and function. Oikos. 1990;**57**: 25-41. DOI: 10.2307/3565733

[57] Barbour MT, Gerritsen J, Snyder BD, Stribling JB. Rapid Bioassessment Protocols for Use in Streams and Wadeable Rivers: Periphyton, Benthic Macroinvertebrates and Fish. 2nd ed. Washington, DC: Report EPA 841-B-99-002, US Environmental Protection Agency, Office of Water; 1999. p. 337

[58] Weisberg SB, Ranasinghe JA, Dauer DM, Schaffner LC, Diaz RJ, Frithsen JB. An estuarine benthic index of biotic integrity (B-IBI) for Chesapeake Bay. Estuaries. 1997;**20**(1): 149-158. DOI: 10.2307/1352728

[59] Engle VD, Summers JK, Gaston GR. A benthic index of environmental condition of Gulf of Mexico estuaries. Estuaries. 1994;**17**:372-384. DOI: 10.2307/1352670

[60] Paul JF, Scott KJ, Campbell DE, Gentile JH, Strobel CS, Valente RM, et al. Developing and applying a benthic index of estuarine condition for the Virginian biogeographic province. Ecological Indicators. 2001;**1**:83-99. DOI: 10.1016/S1470-160X(01)00010-3

[61] Stoddard JL, Herlihy AT, Peck DV, Hughes RM, Whittier TR, Tarquinio E. A process for creating multimetric indices for large-scale aquatic surveys. Journal of the North American Benthological Society. 2008;**27**(4):878-891. DOI: 10.1899/08-053.1

[62] Ganasan V, Hughes RM. Application of an index of biological integrity (IBI) to fish assemblages of the rivers Khan and Kshipra (Madhya Pradesh), India. Freshwater Biology. 1998;**40**:367-383. DOI: 10.1046/j.1365-2427.1998.00347.x

[63] Blocksom KA, Kurtenbach JP, Klemm DJ, Fulk FA, Cormier SM. Development and evaluation of the Lake macroinvertebrate integrity Inde (LMII) for New Jersey lakes and reservoirs. Environment Monitoring and Assessment. 2002;**77**:311-333. DOI: 10.1023/A:1016096925401

[64] Weisberg SB, Thompson B, Ranasinghe JA, Montagne DE, Cadien DB, Dauer DM, et al. The level of agreement among experts applying best professional judgment to assess the condition of benthic in faunal communities. Ecology Indicators. 2008; **8**:389-394. DOI: 10.1016/j.ecolind.2007.04.001

[65] Clarke RT, Wright JF, Furse MT. RIVPACS models for predicting the expected macroinvertebrate fauna and assessing the ecological quality of rivers.

Ecological Modelling. 2003;**160**(3): 219-233. DOI: 10.1016/S0304-3800(02) 00255-7

[66] Davy-Bowker J, Clarke RT, Johnson RK, Kokes J, Murphy JF, Zahradkova S. A comparison of the European Water Framework Directive physical typology and RIVPACS-type models as alternative methods of establishing reference conditions for benthic macroinvertebrates. Hydrobiologia. 2006;**566**:91-105. DOI: 10.1007/s10750-006-0068-5

[67] Horton RK. An index number system for rating water quality. Journal of Water Pollution Control Federation. 1965;**37**(3):300-306. DOI: 10.4236/jep.2013.48099

[68] Tampo L. Hydrochimie et hydrobiologie du bassin du Zio (Togo). Editions Universitaires Européenne. 2018. p. 268

[69] Prati L, Pavanello R, Pesarin F. Assessment of surface water quality by a single index of pollution. Water Research. 1971;**5**:741-751. DOI: 10.1016/0043-1354(71)90097-2

[70] Nguyen HH, Everaert G, Gabriels W, Hoang TH, Goethals PLM. A multimetric macroinvertebrate index for assessing the water quality of the Cau river basin in Vietnam. Limnologica. 2014;**45**:16-23. DOI: 10.1016/j.limno.2013.10.001

[71] Adelagun ROA, Etim EE, Godwin OE. Application of water quality index for the assessment of water from different sources in Nigeria. In: Moujdin IA, Summers JK, editors. Promising Techniques for Wastewater Treatment and Water Quality Assessment. London, UK: IntechOpen; 2021. pp. 1-19. DOI: 10.5772/intechopen.87732

[72] Iwar RT, Utsev JT, Hassan M. Assessment of heavy metal and physico-chemical pollution loadings of river Benue water at Makurdi using water quality index (WQI) and multivariate statistics. Applied Water Science. 2021;**11**:124. DOI: 10.1007/s13201-021-01456-8

[73] Canadian Council of Ministers of Environment (CCME). Water Quality Index User's Manual. Canadian Water Quality Guidelines for the Protection of Aquatic Life; 2001. pp. 1-5

[74] Etim EE, Odoh R, Itodo AU, Umoh SD, Lawal U. Water quality index for the assessment of water quality from different sources in the Niger Delta region of Nigeria. Frontiers in Science. 2013;**3**:89-95. DOI: 10.5923/j.fs.20130303.02

[75] Darko HF, Ansa-Asare O, Paintsil A. A number description of Ghanaian water quality—A case study of the southwestern and coastal rivers systems of Ghana. Journal of Environmental Protection. 2013;**4**:1318-1327. DOI: 10.4236/jep.2013.411153

[76] Miyittah MK, Tulashie SK, Tsyawo FW, Sarfo JK, Darko AA. Assessment of surface water quality status of the Aby Lagoon System in the Western region of Ghana. Heliyon. 2020; **6**:e04466. DOI: 10.1016/j.heliyon.2020.e04466

[77] Banoeng-Yakubo B, Yidana SM, Nti ET, Akabzaa Asiedu D. Analysis of groundwater quality using water quality index and conventional graphical methods: The Volta region, Ghana. Environ Earth Science. 2009;**59**:867. DOI: 10.1007/s12665-009-0082-9

[78] Boateng TW, Opoku F, Osafo Acquaah S, Akoto O. Groundwater quality assessment using statistical

approach and water quality index in Ejisu-Juaben municipality, Ghana. Environment and Earth Science. 2016;**75**: 489. DOI: 10.1007/s12665-015-5105-0

[79] Napo G, Akpataku KV, Alfa-Sika Mande SL, Gnazou MDT, Bawa LM, Djaneye – Boundjou G. Assessment of shallow groundwater quality using water quality index and human risk assessment in the Vogan-Attitogon plateau, Southeastern (Togo). Journal of Environment Pollution and Human Health. 2021;**9**(2):50-63. DOI: 10.12691/jephh-9-2-4

Chapter 5

Water Quality Effects on Growth and Survival of *Oreochromis jipe* and *Oreochromis niloticus* Species in Aquaculture

Job O. Omweno, Albert Getabu, Reuben Omondi and Paul S. Orina

Abstract

The study was conducted at Kegati Aquaculture Center (KMFRI) for 12 weeks to evaluate the effect of water quality parameters on the growth and survival rates of *Oreochromis jipe* and *Oreochromis niloticus*. The fingerlings (n = 270), were stocked in triplicates in $9m^2$ raised ponds arranged in double series and fed on a 30% CP ration twice a day at 10% total bodyweight. Water quality parameters were measured daily using the YSI multi-parameter meter. Fish total length and bodyweight was determined biweekly using a measuring board and a digital weighing balance, respectively, and survivals were determined from the initial and final counts of fingerlings. Data were analyzed with an unpaired sample t-test using R-software and statistical significance was considered at $p < 0.05$. Temperature, total dissolved solids (TDS), and salinity showed no significant differences, whereas the pH and dissolved oxygen (DO) levels were significantly different ($p < 0.05$) between the *O. jipe* and *O. niloticus*. Furthermore, *O. jipe* attained a lower final mean weight (12.16 ± 0.34 g) compared to *O. niloticus* (29.79 ± 0.91 g). The study recommends a further study be conducted in a controlled culture environment to establish optimal conditions for *O. jipe* culture.

Keywords: water quality parameters, fingerlings, growth in aquaculture, *Oreochromis jipe*, *Oreochromis niloticus*

1. Introduction

Water quality encompasses all environmental and, to some extent, biological factors affecting fish growth, survival, and ecology. The most commonly monitored physicochemical parameters include temperature, pH, turbidity, dissolved oxygen concentration (DO), total dissolved solids (TDS), electrical conductivity, and total suspended solids [1]. Temperature controls dissolved oxygen (DO), pH levels, and chemical processes in water. Optimum temperature speeds up fish metabolic rates, which in turn increases feed consumption to meet the energy demands, while high temperature and pH increase the ammonia toxicity to aquatic organisms [2, 3]. Temperature also increases the rate of decomposition and affects feed digestion and assimilation efficiencies in living organisms. The optimum temperatures range for warm water fish (20–30°C) results in maximum growth and survival of these organisms [4]. The lower and upper lethal limits for tilapia are 10–11°C and 37–38°C, respectively [5].

Water pH and DO concentrations regulate the general metabolism and many physiological processes in the cultured fish, such as nitrite and ammonia toxicity. Most aquatic organisms perform optimally in DO levels ranging between 5 and 9 mg L^{-1}, while the DO concentrations below 3 mg L^{-1} and above 9 mg L^{-1} are detrimental to aquatic life [6]. Although tilapias are highly tolerant to low DO levels, *O. niloticus* can survive short-term exposure to low DO levels of 0.1 mg L^{-1} [7], the optimum performance of the species only occurs at DO concentrations ranging between 4.2 and 5.9 mg L^{-1}.

Tilapia have been reported to tolerate a range of up to 3.7 and 11.0 pH but the recommended optimum pH range for most tilapia species is 6.5 to 9.0 [7, 8]. Deviations above or below this optimum range for the species are likely to bring about behavioral and physiological adjustments, which affect the growth performance and survival as fish try to adapt to stressful pH conditions. The highest and lowest lethal pH limits for most fish species are 3.7 and 11.0 respectively [8]. Low pH decreases the egg hatchability rates, while massive fish mortalities occur due to prolonged exposure to toxic un-ionized ammonia whose concentration increases to concentrations greater than 50% when the pH is greater than nine. Ammonia concentrations higher than 0.2 mg L^{-1} are detrimental to fish and are known to depress the appetite of tilapia [5].

2. Materials and methods

2.1 Study area

The study experiment was conducted for 12 weeks from September to December 2019 at Kenya Marine and Fisheries Research Institute (KMFRI), Kegati Aquaculture Center.

Kegati Aquaculture Center is located in a high-altitude area (1974 m asl) between latitudes and longitudes 00420 50.44'S and 0344 470 59.4'E as shown in **Figure 1**. The center receives an averagely high rainfall amount of 1800 ± 100 mm per annum and has a mean temperature range of 20.3–23.9°C. It is drained by river Kuja with a total catchment area of 5180 km^2 [9]. The sampling stations were selected based on accessibility, availability of a well-equipped hatchery facility, and technical support for the study.

Six raised wooden ponds measuring $9m^2$ in a double series and a regular arrangement were randomly assigned to the two treatments in three replicates. The ponds were filled with screened un-chlorinated spring water through an independent inlet to 0.6 m level and water exchange was provided through an outlet throughout the study to promote aeration and siphoning of wastewater containing feed remnants, bioflocs, and fecal waste at 4 days interval. The mixed-sex *O. niloticus* and *O. jipe* fingerlings with an initial mean weight of 2.93 ± 0.12 g and 2.69 ± 0.10 g, respectively, were stocked in the raised wooden ponds at a stocking density of 5 fish m-3.

The fingerlings were fed on 1 mm commercial floating pellets with 30% CP. The daily ration was divided into two equal portions, which were fed regularly to fish twice a day between 0930 h and 1500 h EAT time throughout the study period

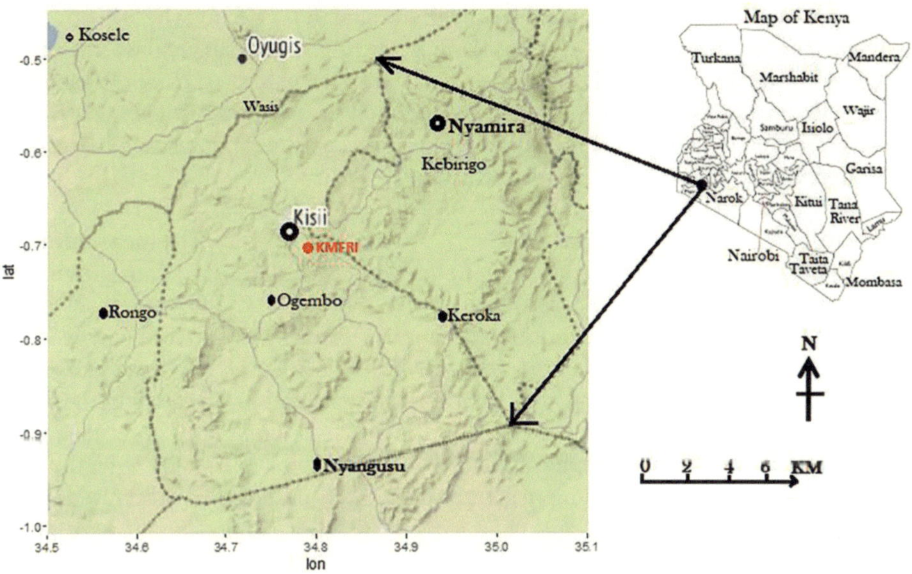

Figure 1.
Map of the study area at KMFRI – Kegati aquaculture research Center in Kisii County, Kenya.

as recommended by ref. [10]. Fish feed administration was done manually which enabled regular inspection of the individual fish and close monitoring of the feeding process, and identifying any dead fish in the pond, which was immediately removed, and the number of mortalities recorded. The feeding rate was at 10% per calculated bodyweight of all fish but the daily ration was divided into two halves for feeding the fish in the morning and afternoon.

2.2 Data collection

Fish growth performance was assessed using average and cumulative bodyweights, which were taken progressively on fingerlings sampled from each raised pond at the end of every 2 weeks. A digital electronic balance (TX 4202 L, SHIMADZU - Corporation, Philippines) with a precision level of 0.01 g was used to record bodyweights. The fish were starved for 24 hours before sampling. Sampling was done very early in the morning before 0900 h to reduce handling stress, which can cause mortality of fingerlings. A sample of 30 fish per tank was taken randomly from each raised pond using a 0.5-inch scoop net. The fingerlings were weighed and returned to their respective ponds. The fingerlings were blot-dried using filter paper before weighing as recommended by ref. [11] and the measurements were taken inside the hatchery to avoid the effects of wind on the digital weighing balance [12]. The fish total length was measured using a graduated board with a readability of 0.1 cm.

After taking both total length and bodyweight measurements, the fingerlings were immediately returned to their respective labeled raised wooden ponds and no feeding was administered until the fingerlings completely recovered from handling stress and were observed freely swimming in the pond. Any sick fish noticed was removed and given prophylactic treatment using 5% sodium chloride solution before returning it to the pond. This was when feeding was administered according to the experimental design. The progressive length-weight measurements were taken biweekly inside the hatchery to avoid the effects of wind on the digital balance [12] and the adjustment of the feeding rates for the fingerlings was done accordingly. Fish survival was determined by counting all stocked fingerlings in each replicate at the start and the end of the growth experiment. After recording the total length and bodyweight, the fingerlings were restocked in their respective culture ponds and feeding resumed after they were observed to have recovered from stress.

The physicochemical parameters determined during the study were DO (mg L^{-1}), temperature (°C), pH, conductivity (μS cm^{-1}), TDS, and salinity. The measurements were regularly taken during the morning hours between 10 am and 11 am throughout the week before the fish were fed their normal daily ration. The water level was maintained at 0.6 m and all previous feed remains were removed before new feeding was administered. Water quality variables: temperature, pH, dissolved oxygen (DO), salinity, conductivity, and total dissolved solids were recorded using the YSI multi-parameter meter (H9829 model), by taking the readings thrice and finding the average. The samples for water physicochemical parameters were collected daily in triplicates and analyzed according to standard methods described by the American Public Health Association [13].

2.3 Data analysis

The collected data were tested for normality using the Shapiro–Wilk test [14], and homogeneity of variances using Bartlett's test. Data were normally distributed and variances homogenously spread, therefore, an unpaired sample t-test was used to compare the population means of *O. jipe* and *O. niloticus* treatments. The variations in observed datasets on growth performance, survival rates, and the water quality variables were compared between *O. jipe* and *O. niloticus* treatments using robust mean ± standard error and range and nonrobust mean ± standard deviation using the R-software programming procedures [15]. All data analyses were performed using the 64-bit R-software version 3.6.3 [15], and the observed differences were considered statistically significant at $p < 0.05$.

2.4 Determination of growth and survival rates

The fish growth and survival parameters were computed with the following equations according to [10]:

$$\text{Specific growth rate}\left(SGR\right)\%/day = \frac{\boldsymbol{In\,Wt - In\,W0}}{\boldsymbol{t}} \times \mathbf{100} \tag{1}$$

$$\text{Feed conversion ratio}\left(FCR\right) = \frac{\boldsymbol{Consumed\ feed\left(g\right)}}{\mathbf{W1 - W0}} \tag{2}$$

where W0 and W1 are the initial fish weight and the final fish weight (g) of the fingerlings, in is the natural logarithm, and t is the number of culture days.

$$\text{Survival rate}\left(\%\right) = \frac{\boldsymbol{final\ fish\ count}}{\boldsymbol{initialfish\ count}} \times \mathbf{100} \tag{3}$$

$$Net \text{ yield}\left(kg\ \text{m}^{-3}\right) = \text{Harvest biomass} - \text{Stocking biomass} \quad (4)$$

3. Results

3.1 Fish growth, survival, and production parameters

The general growth trend showed that there was a steady gain in bodyweight in both *O. jipe* and *O. niloticus* during the study period. The un-robust comparisons of combined boxplots and standard deviations show a wide variation in data recorded from the two treatments. The mean, standard deviations, and variance of fish body-weights in *O. niloticus* and *O. jipe* exhibited a wide range of variation and increased exponentially during the culture period (**Figure 2**). The means and standard error variations for the growth, survival, and production indices of *O. jipe* and *O. niloticus* treatments are shown in **Table 1**.

3.1.1 Survival rates

Both species recorded high survival rates ranging between 74.6% and 95% in *O. jipe* treatment, while in *O. niloticus*, it ranged between 94.6% and 100%. Hence, the mean survival rate in *O. jipe* (92 ± 1.19%) was significantly lower ($p < 0.05$) than the mean survival rate (96.28 ± 0.65%) of *O. niloticus* treatment. Whereas, *O. niloticus* maintained high survival rates throughout the culture period, the lowest survival rates were recorded in *O. jipe* treatment due to mortalities that occurred during the culture period.

3.1.2 Growth parameters

The observed fish growth in terms of bodyweight indicated both species were initially growing slowly during the first few (1–3) weeks of the culture of the study

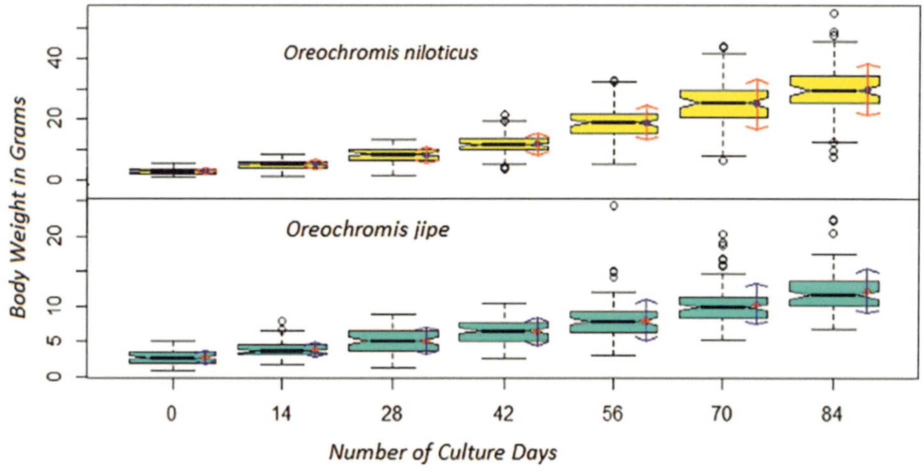

Figure 2.
Data variation in Oreochromis niloticus *and* Oreochromis jipe *using a comparison of boxplots and non-robust mean +/− SD.*

Parameter	O. niloticus		O. jipe		n	t - test sig.
	Mean ± SE	Range	mean	Range		
Initial length (cm)	4.80 ± 0.39	2.4–7.1	4.67 ± 0.17	1.82–7.85	180	p = 0.23
Final length (cm)	11.51 ± 0.33	8.1–16.7	9.11 ± 0.72	6.5–12.0	179	p = 0.01
Initial weight (g)	2.93 ± 0.25	0.61–5.64	2.69 ± 0.20	0.82–5.05	180	p = 0.14
Final weight (g)	29.79 ± 0.91	17.7– 55.2	12.16 ± 1.34	6.78–22.39	179	p < 0.05
SGR (gfish^{-1} day^{-1})	2.28 ± 0.21	0.36–3.76	1.45 ± 0.74	0.12–4.74	170	p < 0.05
%Weight Gain (gfish^{-1}).	4.32 ± 0.32	2.51–5.67	1.58 ± 0.35	0.98–2.73	170	p < 0.05
DWG (gfish^{-1} day^{-1})	0.32 ± 0.17	0.11–0.96	0.113 ± 0.01	0.07–0.17	170	p < 0.05
FCR	1.59 ± 0.12	1.21–1.68	2.502 ± 0.15	2.35–2.57	12	p = 0.02
Survival rates (%)	96.28 ± 0.95	94.6–100	92 ± 1.19	74.6–95.0	35	p = 0.02
Fish yield (kgm^{-3})	2.59 ± 0.05	2.79–3.15	0.85 ± 0.03	0.79–0.99	170	p < 0.001

Table 1.
Mean (± standard error) of growth, survival and production parameters for Oreochromis jipe *and* Oreochromis niloticus.

but increased exponentially after the second month. The highest and lowest fish total lengths observed during the final sampling were (16.7 cm and 6.5 cm) in *O. niloticus* and *O. jipe*, respectively. However, despite this wide range of variation observed in the total length measurements, the two-sample t-test indicated no significant growth difference ($p > 0.05$) between the mean final length (TL = 11.51 ± 0.13 cm) of *O. niloticus* and *O. jipe* (TL = 9.11 ± 0.12 cm).

The mean growth in final bodyweight in *O. jipe* (12.18 ± 35 g) was significantly lower ($t = 2.42$, $p < 0.05$) than the mean final weight (29.79 ± 0.91 g) for *O. niloticus* (**Figure 3**). Similarly, there was a significant difference ($t = 2.57$, $p < 0.05$) in the SGR between the two species; *O. jipe* had a lower mean SGR (1.45 ± 0.14 g/fish^{-1} day^{-1}) in comparison with (2.28 ± 0.26 g fish^{-1} day^{-1}) O. 3 *niloticus*. The DWG also showed the same trend in both species. The DWG for *O. niloticus* (0.32 ± 0.17 g fish^{-1} day^{-1}) was significantly higher ($t = 4.62$, $p < 0.05$), than that of *O. jipe* (0.1128 ± 0.01 g fish^{-1} day^{-1}). The %weight gain (WG) of *O. jipe* (1.58 ± 0.35 g fish-1) was significantly lower, ($t = 3.57$, $P < 0.05$) compared to the mean weight gain of *O. niloticus* (4.48 ± 1.6 g fish^{-1}).

3.1.3 Production parameters

The mean fish yield of *O. jipe* species was significantly lower ($t = 4.46$, $p < 0.05$) in *O. jipe* compared to the *O. niloticus* species. The yield in *O. jipe* ranged between 0.79 and 0.99 kg m^{-3}, while in *O. niloticus* it ranged between 2.79 and 3.15 kg m^{-3}. Also, there was a significant difference ($t = 2.5$, $p < 0.05$) in FCR between *O. jipe* and *O. niloticus* species. The mean FCR varied between 2.35 and 2.57, whereas in *O. niloticus* it ranged between 1.21 and 1.68 during the culture period. Furthermore, the reciprocal of FCR (FCR^{-1}) yielded food conversion efficiencies (FCEs) for *O. niloticus,* which were significantly higher than those of *O. jipe* species.

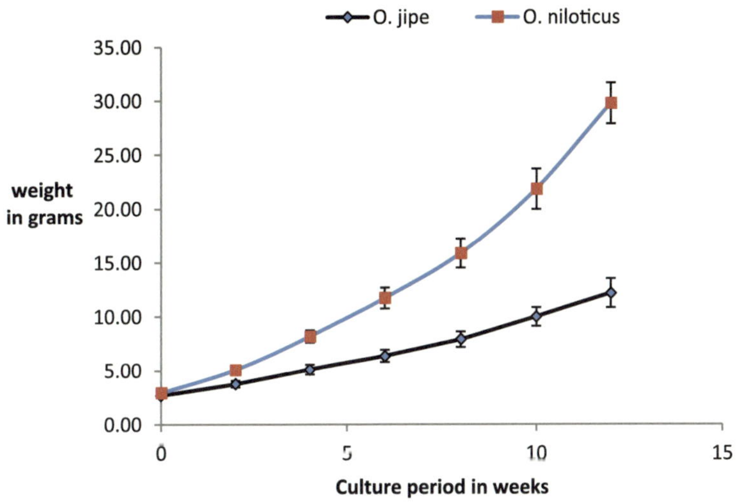

Figure 3.
Weight gain in cultured Oreochromis jipe and Oreochromis niloticus.

3.1.4 Water quality parameters

The mean values and variation of water quality parameters recorded during the study are shown in **Table 2**.

Water temperature showed no significant difference ($t = 0.14$, $p > 0.05$) between *O. jipe* treatment and the control during the culture period. The highest mean temperature recorded was 24.3 ± 0.58°C in *the O. jipe* culture system, while the temperature in *O. niloticus* treatment ranged between 18.5 and 24.8°C, with a mean of 22.54 ± 0.39°C. However, the DO level was significantly higher ($t_{129} = 4.31$, $p < 0.05$) in the control, whereas the pH was significantly ($t129 = 2.91$, $p < 0.05$) lower in *O. jipe* treatment than *O. niloticus* treatment. The water pH of both culture systems ranged between 7.34 ± 0.09 and 8.37 ± 0.14, whereas DO concentrations ranged between 4.9 ± 0.20 mg L^{-1} and 6.75 ± 0.14 mg L^{-1}. Total dissolved solids (TDS), conductivity,

	O. niloticus		***O. jipe***			**Unpaired t-test**
Parameter	**Mean (± S.E)**	**Range**	**Mean (± S.E)**	**Range**	**N**	
Temp.	22.54 ± 0.39	18.5–24.8	22.93 ± 0.31	20.9–26.5	130	p = 0.089
DO	5.94 ± 0.10	4.99 – 6.94	5.04 ± 0.11	4.54–9.37	130	p < 0.001
pH	7.54 ± 0.09	6.88–8.69	8.01 ± 0.13	6.54–9.37	130	p = 0.004
TDS	41.39 ± 3.51	31.6–57.5	40.42 ± 3.53	32.4–56.8	130	p = 0.143
Salinity	0.03 ± 0.003	0.02–0.04	0.03 ± 0.003	0.02–0.04	130	p = 0.173
EC	57.92 ± 4.61	39.7–92.5	57.2 ± 4.55	31.6–89.9	130	p = 0.520

Table 2.
Mean (± S.E) of water quality parameters recorded for Oreochromis niloticus *and* Oreochromis jipe *treatments.*

and salinity showed no significant differences ($p > 0.05$) between the treatment and the control. Specifically, TDS and conductivity increased progressively after replenishing water in the experimental ponds until the water was exchanged again. However, lower values of conductivity, TDS, and salinity were recorded during the study. Low salinity could be a result of dilution during the rains and the frequent replenishment of water.

4. Discussion

The most crucial interpretations of the study findings based on the overall objective are presented.

4.1 Data variation in *O. jipe* and *O. niloticus* culture treatments

The overall variation of the data set is greater for both the *O. jipe* and *O. niloticus* species was due to the outliers which ranged from 5 to 20 standard deviations above or below the mean of zero determined by the Partial Least Squares (PLS) regression model [16]. These were attributed to a small number of the faster-growing individuals (shooters), which gained weight faster than the rest of the mixed-sex fingerlings, resulting in un-uniform sizes of fish in each treatment group.

4.2 Growth and survival rates

The study used uniformly homogenous fingerlings, which were expected to grow uniformly throughout the culture period. Upon release into the pond, the surviving fingerlings of both species displayed good condition and consumed feed during the 12-hour monitoring period. However, both growth performance and survival rates of *O. jipe* were significantly lower than those of *O. niloticus.* This would have been due to the faster adaptability of *O. niloticus* to the aquaculture environment because of the previously reported good culture attributes [17]. Chenyambuga et al. [18], confirm that *O. niloticus* quickly adapts to aquaculture conditions in a comparative study in which the exotic Nile tilapia, *O. niloticus* attained a final mean weight of 67.6 ± 2.4 g and SGR of 2.2 ± 0.14 g $fish^{-1}$ day^{-1}, outperforming *O. jipe* with a final mean weight of 16.3 ± 2.0 and SGR of 1.5 g $fish^{-1}$ day^{-1} attained within the same culture period of 90 days. This study however reported lower values of final mean weight and SGR in *O. jipe,* probably due to temperature difference between the two regions. Whereas, [18] recorded a mean temperature of 25.2 ± 2.0°C, the present study recorded a lower mean temperature of 22.93 ± 0.31°C for the *O. jipe* species. This temperature lies below the range of 20–30°C recommended for optimum growth and survival of most tilapia species [19, 20]. Nevertheless, the findings on survival rates of *O. jipe* and *O. niloticus* agree with the study by [18], who reported higher survival rates of 95.8% in *O. jipe* and 100% survival in *O. niloticus*, suggesting that both species can have exceptionally high survival rates if properly managed in the culture system. The findings on survival also concur with Hussain et al. [21], who reported 100% survival of *O. niloticus* in a polyculture experiment. These findings on survival however differ from the results of [22], who reported low survival rates of 25.8% for *O. jipe.* The low survival rates of *O. jipe* reported were attributed to decomposing feed remnants (bioflocs) which were not properly managed in the culture system during the culture trials.

The present study used an open pond system that is semi-controlled by natural factors and the siphoning of bioflocs was done before they settled and decomposed at the pond bottom. This high level of management ensures that the initially slow-growing fish attained a steady and exponential increase in total length and bodyweight, which was recorded from the second month onwards. As a result of slow acclimatization to the captive environment, both species exhibited nonoptimal performance [23]. Low DO levels have been reported to cause fish mortalities and other challenges in the aquaculture production system. The amount of dissolved oxygen consumed by the fish depends on fish size, feeding rate, activity level, and water temperature. In this study, the *O. jipe* species displayed a remarkable reduction in feed intake immediately after the rains, which might have resulted in the observed low DO levels in this culture treatment. The same phenomenon has been reported in other cultured fish, such as *Labeo victorianus* by [24], in which the accumulation of unconsumed feed remnants resulted in prolonged low DO levels in the culture ponds, which stressed the fish, resulting in decreased growth and survival rates. Several studies have reported a decline in fish weight gain which has been attributed to decreased water temperatures during cold seasons characterized which directly affect feed intake [21, 25]. This might have occurred in the present study where water physicochemical parameters were not controlled although the pH and DO levels were within the recommended range for tilapia growth [26].

4.3 Water quality parameters

4.3.1 Water temperature

The narrow range of water temperature in both *O. niloticus* and *O. jipe* species could be due to the time that temperature readings were taken during the morning hours, which was between 9.00 am and 11.00 am daily. The study, however, did not record any lethal temperature values which range between 10–11°C and 37–38°C, respectively [4]. Although the mean temperatures for both treatments were within the optimum range of 20–30°C required for fish growth [4, 5, 27], these temperatures were below the specific optimum range of 25–30°C recommended for Tilapia growth. The low temperatures could be attributed to a high altitude location of the study area with low mean ambient temperatures ranging between 18°C and 26°C. Although water temperature did not show any significant differences between the *O. jipe* and *O. niloticus* treatments, it has been reported that water temperatures below the range of 20–22°C contribute to a nearly 30% decline in optimal growth [5]. This could be the reason for the nonoptimal growth performance of *O. jipe* and *O. niloticus* in aquaculture. Probably, the novel species *O. jipe* were more stressed due to slow adaptability to the changing physicochemical parameters than *O. niloticus*.

4.3.2 Dissolved oxygen (DO) concentration

The reduction of DO levels in *the O. jipe* culture system was probably due to low acceptance of commercial feeds, which were administered in their daily ration. *O. niloticus* species on the other hand consumed all the feed provided within a span of 2–3 hours. This necessitated frequent siphoning of feed remnants that settled at the bottom of *O. jipe* ponds and flushing of the culture systems with clean fresh water from the tanks was done uniformly to all ponds was done every 2 days. Although frequent siphoning was done regularly, the lack of efficient removal of all bioflocs contributed to the deterioration of water quality resulting in significant differences in DO levels between the *O. jipe* and *O. niloticus* culture systems. This result on *O. jipe* DO levels corroborates with the findings of [22] that attributed low to feed remnants that settled at the bottom of the hapa-in-pond culture system. Lower DO levels have been reported to decrease feeding and respiration in most tilapia species and make it difficult for fish to assimilate the consumed feed [5, 28]. In addition, the amount of dissolved oxygen in the raised ponds was insufficient because the small-sized fingerlings tend to consume additional oxygen to meet the demands for their increased metabolic rates, which doubles for every 10 degrees increase in water temperature within the optimum range [29].

4.3.3 pH

The study found that the mean pH (8.01 ± 0.13) of *O. jipe* treatment was significantly higher than the mean pH (7.54 ± 0.09) of *O. niloticus.* Probably, unconsumed pellets particularly in *O. jipe* treatment could have been decomposed under low DO levels to liberate ammonia (NH_3), which accounts for the range of pH values greater than nine. Although ammonia levels were not determined during the study, studies have shown that ammonia is derived from the breakdown of nontoxic ammonium ion ($NH4^+$) contained in digestible crude protein and is a component of total nitrogen highly toxic to fish. However, good water management ensured that the pH did not affect the fish growth and survival because it prevented prolonged exposure of fish to toxic un-ionized ammonia concentrations of greater than 0.2 mg L^{-1}, which can depress the appetite of Tilapia fish [5]. This partly explains why the pH was not consistently high during the study but showed fluctuations between high and low values.

5. Conclusions and recommendation

The study findings showed that the investigated water quality parameters affected the growth performance and survival rates of *O. jipe* and *O. niloticus* cultured using raised wooden ponds. As a result, both species did not realize their optimal performance in the aquaculture environment because the temperature fell below the range of 25–30°C recommended for tilapia culture. This might have contributed to low food intake by the species, resulting in low DO and high pH levels. Fish growth was however influenced by the interplay of unmonitored physicochemical parameters in an open pond system. However, for the successful introduction of *O. jipe* species to aquaculture and optimal growth in the aquaculture environment, it is crucial to delineate optimal culture conditions for the species. This can be done by monitoring the effect of water quality parameters on the growth and survival of fish in a controlled fish culture system and monitoring growth for a longer period. The necessity for a repeat of this experiment in a different culture system, such as hatchery tanks or a greenhouse, to determine whether the results obtained can be replicated under different cultural conditions. In addition, water quality can be recorded using automatic sensor detectors, which take the readings over infinitesimal time intervals and the data collected can be modeled using advanced differential models to evaluate several latent variables in the underlying relationships between water quality and growth parameters.

Acknowledgements

The authors wish to acknowledge the African Development Bank for funding this research through an MSc Scholarship awarded to Mr. J. O. Omweno. Thanks to the Kenya Marine and Fisheries Research Institute (KMFRI), and Kegati Aquaculture Center for technical support.

Conflict of interest

The authors declare that they have no competing personal or professional interests. Further, the funding agency did not play any role in the conceptualization, design, and implementation of the study, as well as the decision to publish the results.

Ethics, approval, and compliance statement

The study was conducted by following all the applicable guidelines stipulated by the National Commission for Science, Technology, and Innovation (NACOSTI) regulations (2014) under the research permit no NACOSTI/P/20/3982 and the Kenya Marine and Fisheries Research Institute (KMFRI) set by Cap 250 of the Science and Technology Act (1979), which has since been repealed by the Science, Technology and Innovation Act (Act no. 28, 2013) and Section 56 of the act that mandates KMFRI to research marine and freshwater resources.

Author details

Job O. Omweno[1*], Albert Getabu[1], Reuben Omondi[1] and Paul S. Orina[2]

1 Department of Environment, Natural Resources and Aquatic Sciences, Kisii University, Kisii, Kenya

2 Kenya Marine and Fisheries Research Institute (KMFRI), Kegati Aquaculture Centre, Kisii, Kenya

*Address all correspondence to: omwenojob@gmail.com

References

[1] Marshall S, Elliott M. Environmental influences on the fish assemblage of the Humber estuary, U.K. Estuarine. Coastal and Shelf. Science. 1998;**46**:175-184

[2] United States Environmental Protection Agency (USEPA). Aquatic Life Ambient Water Quality Criteria for Ammonia-Freshwater. USEPA; 2013. pp. 8-22. Available from: http://www.epa.gov/safewater/consumer/2ndstandards.html. [Accessed: April 09, 2022]

[3] Shackleton E. Effects of temperature and terrestrial carbon on fish growth and pelagic food web efficiency. PhD Thesis. Sweden: Umeå University; 2012. Available from: http://umu.diva-portal.org

[4] Mirea ET. Influence of Different Water Temperature on Intensive Growth Performance of Nile Tilapia in a Recirculation Aquaculture System. New York, USA: CRC Press; 2013. pp. 65-97

[5] Nehemia A, Maganira JD, Rumisha C. Length-weight relationship and condition factor of Tilapia species grown in marine and fresh water ponds. Agriculture Biolog. 2012;**3**(3):117-124

[6] USEPA. Secondary drinking water Regulations: Guidance for Nuisance Chemicals. 1992. Available from: http://www.epa.gov/safewater/consumer/2ndstandards.html. [Accessed: September 12, 2020]

[7] Ghozlan AA, Zaki MM, Gaber MM, Nour A. Effect of different water sources on survival rate (%) growth performance, feed utilization, fish yield, and economic evaluation on Nile Tilapia (*Oreochromis niloticus*) Monosex reared in earthen ponds. Oceanography Fish Open Access Journal. 2017;**6**(1):555-576. DOI: 10.19080/OFOAJ.2018.05.555676002

[8] Omer NH. Water Quality Parameters. In: Summers K, editor. Water Quality - Science, Assessments and Policy [Internet]. London: IntechOpen; 2019. DOI: 10.5772/intechopen.89657. [cited 2022 Aug 13]. Available from: https://www.intechopen.com/chapters/69568

[9] Ojany FF, Ogendo RB. Kenya: A Study in Physical and Human Geography; 1973. Kenya: University of Nairobi Research Archives. 1973. p. 228. Available from: http://erepository.uonbi.ac.ke

[10] Thongprajukaew K, Kovitvadhi S, Kovitvadhi U, Preprame P. Effects of feeding frequency on growth performance and digestive enzyme activity of sex-reversed Nile tilapia, Oreochromis niloticus (Linnaeus, 1758). Agriculture and Natural Resources. 2017;**51**:292-298. DOI: 10.1016/j.anres.2017.04.005

[11] Anderson RO, Gutreuter SJ. Length, weight and associated structural indices. In: Nielsen LA, Johnson DL, editors. Fisheries Techniques. Bethesda, Maryland: American Fish Society; 1983

[12] Shoko AP, Limbu SM, Mrosso HDK, Mkenda AF, Mgaya YD. Reproductive biology of female Nile tilapia *Oreochromis niloticus* (Linneaus) reared in monoculture and polyculture with African sharptooth catfish *Clarias gariepinus* (Buchell). Springerplus. 2015;**4**:275-290

[13] APHA. Standard Methods for the Examination of Water and Wastewater. Washington, DC: American Public Health Association; 1998. p. 1325

[14] Shapiro SS, Wilk MB. An analysis of variance test for normality

(complete samples). Biometrika. 1965;**52**(3/4):91-611

[15] R-core team. R: A Language and Environment for Statistical Computing. Vienna, Austria: R Foundation for Statistical Computing; 2020

[16] Omweno JO, Getabu A, Orina PS, Omasaki SK, Zablon WO. PLS regression to compare the relative importance of water quality variables in fish growth. Journal of Applied Structural Equation Modeling. 2021;**5**(1):1-8

[17] Courtenay WR Jr. Tilapias as non-indigenous species in the Americas: Environmental, regulatory, and legal issues. In: Costa-Pierce BA, Rakocy JE, editors. Tilapia Aquaculture in the Americas. Baton Rouge, Louisiana, United States: World Aquaculture Society; 1997

[18] Chenyambuga SW, Meiluidie E, Madalla N, Mnembuka-Sokoine VB, Lochmann R. Evaluation of Performance of Different Tilapia Species. Quality Seedstock Development/ Experiment/09QSD04PU. Final Reports: Investigations 20092011, 2011

[19] Ngugi CC, Bowman JR, Omolo BO. A new guide to fish farming in Kenya. Aquaculture Collaborative Research Support Program (ACRSP). 2007;**2007**:1-100

[20] Makori AJ, Obuom PO, Kapiyo R, Anyona ND, Dida GO. Effects of water physico-chemical parameters on tilapia (*Oreochromis niloticus*) grown in earthen ponds in Teso north sub-county, Busia County. International Journal of Fisheries and Aquatic Sciences. 2017;**20**(30):1-10

[21] Hussain M, Hussain SM, Mumtaz WM, Abbas NM, Raza A, Mubariki SM. Comparative study to monitor survival and growth of varying fish species ratios under the same stocking density. World Applied Sciences Journal. 2014;**30**(12):1784-1790

[22] Ogada EO, Orina PS, Morara Ogendi GM. Effect of stocking density and diet on the performance of Jipe tilapia (*Oreochromis jipe*) cultured in hapas at Sagana, Kenya. International Journal of Advanced Research. 2018;**6**(10):296-308

[23] Musa S, Orina PS, Aura CM, Kundu R, Ogello EO, Munguti JM. The effects of dietary levels of protein and greenhouse on growth, behaviour and fecundity of Nile tilapia (*Oreochromis niloticus* L.) broodstock. International Journal of Science and Research (IJSR). 2012;**2**(13):2271-2294

[24] Orina PS, Charo-Karisa H, Munguti JM, Boera P, Abwao J, Kyule D, et al. A comparative study of Labeo victorianus (Bouelenger, 1901) and *Oreochromis niloticus* (Linnaeus, 1758) grown in polyculture systems. Lakes & Reservation. 2018;**00**:1-7

[25] Nazish N, Mateen A. Winter growth of carps under different semi-intensive culture conditions. Pakistan Veterinary Journal. 2011;**31**:134-136

[26] Gustavasson H. Locally available protein sources in diets of Nile tilapia(*Oreochromis niloticus*). A study of growth performance in the Mekong Delta in Vietnam. Masters thesis. Sweden: Swedish University of Agricultural Sciences; 2016. 47 p

[27] Mjoun K, Rosentrater K, Broun M. Tilapia: Environmental biology and nutritional requirements. Fact Sheets. 2010;**2010**:164

[28] Mallya YJ. The effects of dissolved oxygen on fish growth in aquaculture. Kingolwira National Fish Farm Center

UNU-Fisheries Training Programme; 2007. Available from: http://www.unuftp.is->yovita07prf.pdf

[29] Gingerich AJ, Cooke SJ, Hanson KC, Donaldson MR, Hasler CT, Suski CD, et al. Evaluation of the interactive effects of air exposure duration and water temperature on the condition and survival of angled and released fish. Fisheries Research. 2007;**86**:169-178

Section 3

Chemical Quality of Surface and Groundwaters

Chapter 6

Bacteriological Perspective of Water Quality

Hatice Aysun Mercimek Takci, Melis Sumengen Ozdenefe, Fikret Buyukkaya Kayis and Sadık Dincer

Abstract

Water quality can be defined as a measurement of a water's appropriateness for a specific purpose based on biological, chemical, and physical qualities. Water pollution caused by microorganisms is one of the most serious threats to the aquatic ecosystem around the world. The bacterial concentration in an aquatic environment is raised by anthropogenic activities and industrial-agricultural pollutants. Coliform bacteria have long been used as an indicator organism for microbial pollution of water, which has contributed to potential health risks. *Escherichia coli* is the coliform that indicates fecal contamination. Various methods such as culture-dependent methods, culture-independent gene sequence-based methods, and immunological methods are used to determine bacterial contamination in water sources. As a consequence, determining that the water is not suitable for use by bacteriological analysis indicates that the water is contaminated. This chapter highlights the sanitary quality of aquatic environments, indicator organisms for water quality assessment, detection of bacterial pollution in the water source, and bacterial waterborne diseases.

Keywords: contamination, *E. coli*, indicator organisms, sanitary, water-borne diseases, water quality

1. Introduction

Water, which is the main constituent of Earth's hydrosphere and one of the most essential natural resources for life, is an inorganic chemical substance [1, 2]. It is consumed by societies for various purposes such as human activities (drinking, household, and recreational), agriculture (crop irrigation and food processing), and industry processes [2, 3]. Water-based environments contain the different communities of higher organisms and microorganisms that interact with each other and their environment [4]. Aquatic ecosystems have two main types: the "marine and freshwater ecosystem" [4, 5]. The largest aquatic ecosystem is marine water, which covers more than 70% of the Earth's surface (estuaries, coral reefs, coastal ecosystems, and oceans). The lotic environment, lentic ecosystem, and wetland habitats are subdivided from freshwater, which constitutes less than 1% of the Earth's aquatic ecosystems and has a lower salt content [5]. The residues of anthropogenic activities (filling and construction of bridges, canals, dams, roads, and deforestation) industrial and agricultural pollutants discharge into aquatic environments [4, 5]. Especially the freshwater bodies are exposed to wastes and leaks such as detergents, heavy metals, plastic or non-plastic origin compounds, microfibers, etc. [6–8]. As well as these chemicals and toxic residues reaching aquatic ecosystems, microbial water pollution that occurs even in developed countries is considered one of the major threats to human health across the globe [2, 9]. WHO has reported approximately 46,000 infant deaths and 600 million cases of diarrhea and dysentery per year as a result of contaminated water and insufficient sanitation [10]. The surface water bodies in contact with domestic or sewage wastes from the surrounding area are potentially hazardous ecosystems as carriers of pathogenic microorganisms [2, 11]. Microbiologic monitoring of water quality is critical in terms of detecting, identifying, and quantifying pathogens that cause waterborne diseases [1]. Various microbiologic and molecular techniques including surveillance, detection methods, analysis, and decision-making processes are used in quantitative microbial risk assessment of possible pathogen contamination [12].

Here, we focus on the sanitary quality of aquatic environments, indicator organisms for water quality assessment, techniques used in monitoring water quality, surveillance, quantitative microbial risk assessment, and bacterial waterborne diseases.

2. Sanitary quality of aquatic environments

The physical, chemical and biological characteristics of water define its quality. There is a lot of diversity in these water quality features all over the world. For this reason, natural water sources' quality in use for various purposes should be determined by means of the water quality parameters [8, 13–15].

Water is classified as surface and groundwater depending on its origin. Agricultural, industrial, and household activities can contaminate both types of water, exposing them to a variety of contaminants such as fertilizers, heavy metals, oils, pesticides, and toxic compounds (**Figure 1**).

Potable water, palatable water, polluted water, and infected water are the four types of water quality.

The following are scientific meanings of these types of water quality:

a. Potable water is water that is safe to drink, tastes well, and can be used in the home.

b. Palatable water has a pleasant esthetic and takes into account the existence of chemicals that are not harmful to the health of humans.

c. Polluted water is inappropriate for drinking or domestic use because it contains undesired physical, chemical, biological, or radioactive components.

d. Infected water is that water containing pathogenic organisms [8].

Physical parameters of water quality include turbidity, temperature, color, taste and odor, solids, and electrical conductivity (EC). The chemical water parameters contain pH, acidity, alkalinity, chloride, chlorine residual, sulfate, nitrogen, fluoride, iron and manganese, copper and zinc, hardness, dissolved oxygen, biochemical oxygen demand (BOD), chemical oxygen demand (COD), toxic inorganic and organic

DOI: http://dx.doi.org/10.5772/intechopen.112874

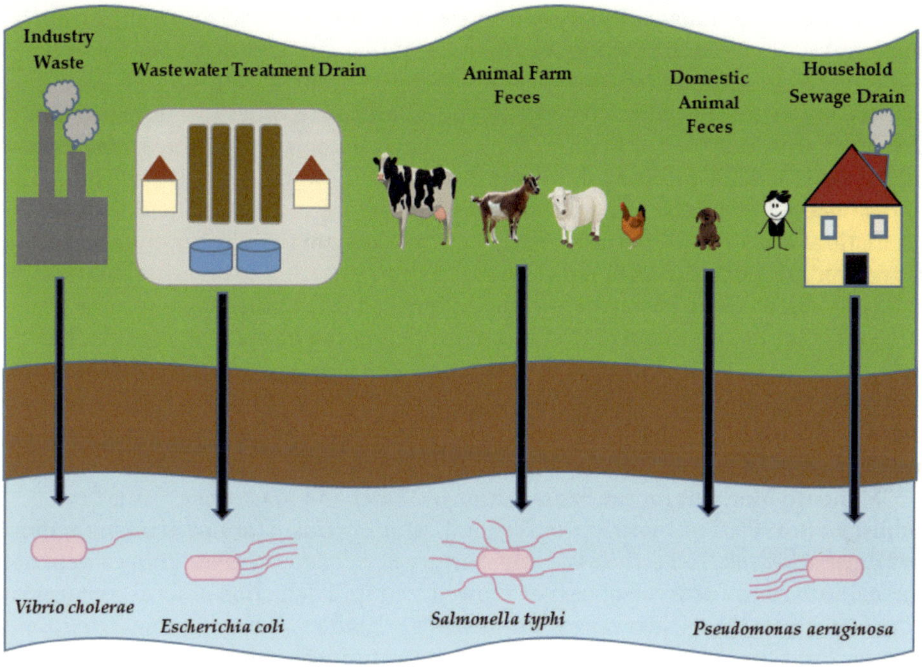

Figure 1.
Various factors affecting of water quality and disease-causing some bacteria [2].

substances, radioactive substances. Water quality is also determined by biological parameters such as bacteria, algae, viruses, and protozoa [2, 8, 13–15]. Considering these parameters, this chapter especially aims to emphasize the bacteriological pollution of water quality.

The presence or absence of live microorganisms can be one of the most beneficial indicators of water quality. Biologists may research organisms in natural waters and measure water quality using an SDI (species diversity index); therefore, a water body with a great number of well-balanced species is considered a healthy environment. The presence of some organisms in aquatic environments is considered an indicator of water pollution [8].

The human intestinal system carries specific population of microorganisms, with coliform bacteria accounting for a large share of this population. Although wastewater contains millions of bacteria per milliliter, the majority of them are safe. The existence of pathogenic bacteria in wastewater can be hazardous when it contains waste from people affected by diseases [8].

3. Indicator organisms for water quality assessment

Nonpathogenic microorganisms which have minimal or no growth in water and can be dependably detected at low concentrations are preferable. Several indicator bacterial species can be utilized in water quality assessment, as explained in the following [3].

Bacteria have been used as indicators of water sanitary quality since 1880 when Von Fritsch characterized *Klebsiella pneumoniae* and *Klebsiella rhinoscleromatis* as microbes commonly found in human feces. Percy and Grace Frankland began the first routine bacteriological analysis of water in London in 1885, counting bacteria using Robert Koch's solid gelatin medium. Escherich also identified *Bacillus coli* from the feces of breast-fed newborns in 1885. The Franklands proposed in 1891 that the organisms found in sewage must be identified in order to obtain proof of possibly hazardous contamination. Sanitary bacteriologists were employing the "Wurtz method" of counting *B. coli* via direct plating water samples on litmus lactose agar by using the principle of acid from lactose as a distinguishing characteristic about 1893. With the invention of the Durham tube, the production of gas proved. In 1901, the term 'coliform' bacterium was used to describe bacteria that resembled *B. coli*. Nevertheless, the colony count for bacteria in water was not officially adopted till Report 71. Through the beginning of the 20th century, bacteriologists accepted the sanitary importance of identifying diverse coliforms, as well as streptococci and *C. perfringens*. MacConkey's broth which was used to identify lactose-fermenting bacteria that were bile salt tolerant was defined by MacConkey in 1905 [16, 17].

"Standard Methods for the Examination of Water and Wastewater" was first published in 1905. The most appropriate indicator organism for raw drinking water was *Escherichia coli*. Nevertheless, early-century *E. coli* detection techniques were not amenable to routine surveillance that allowed for rapid detection of fecal contamination. Consequently, a wider range of organisms related to and containing *E. coli*, dubbed "coliforms" by Blachstein in 1893, were examined as proxy indicators. Since then, the coliform group has been revised depending on the ability to categorize members based on their genetic structure. *Bacterium coli* has been renamed *E. coli*, *Bacterium aerogenes* has been renamed *Enterobacter aerogenes*, and the 'intermediate' varieties have been classed [17].

The discovery that certain 'fecal coliforms' were not of fecal origin, and the development of advanced *E. coli* testing procedures, have led to the use of *E. coli* as the 'favored' indicator for the identification of fecal contamination [17].

The definition of fecal coliforms has been updated to better reflect the genetic structure of its members and to incorporate newly discovered environmental organisms. After all, fecal coliforms are becoming more commonly known as "thermotolerant" coliforms. This, together with enhanced *E. coli* detection techniques, has prompted a tendency to use *E. coli* as a more trustworthy indicator of fecal pollution in drinking water than thermotolerant coliforms.

Escherichia coli, fecal coliforms, and/or Coliforms are the indicator bacterias currently used for drinking water monitoring in developed nations, though the belief of indicator organisms as the primary source of data about the safeness of drinking water is being questioned in so many states. The World Health Organization (WHO) proposed *E. coli* as the primary indicator of fecal contamination in 2003 [17, 18].

Coliforms are all facultative anaerobic, non-spore-forming, gram-negative, rod-shaped bacteria, oxidase-negative, fermenting lactose to acid and gas at 35°C in 48 h or members of the *Enterobacteriaceae* that are β-galactosidase positive [16, 17]. The coliforms contain innocuous *E. coli* and *Enterobacter*, as well as the most prevalent intestinal bacteria and rare pathogens such as *Klebsiella*, *Citrobacter*, *Kluyvera*, and *Leclercia* genera, and also some members of the *Serratia* genus. These microorganisms are utilized as a fecal pollution indicator in water since they are found in the intestinal tracts of homeothermic animals and seem to have sanitary importance [2, 8, 16–19].

Fecal coliforms are bacteria that can grow and ferment lactose with the formation of acid and gas in EC broth (*Escherichia coli* Broth) at 44.5°C within 24 h. This group associated with fecal pollution in warm-blooded animals is known "thermotolerant coliforms". It also includes *E. coli*, is predominantly fecal, as well as other species, which are called non-*E.coli* such as *Klebsiella, Enterobacter*, and *Citrobacter*. Septic systems and sewer, run-off from dairy, feedlot, and farming sites, stormwater, and straightly defecating of livestock to water can introduce the fecal coliforms enter to rivers and streams [3, 8, 13, 16–20].

Fecal streptococci are Gram-positive coccoid bacteria, which were being studied as a key of indicator microorganism contamination. However, difficulties in distinguishing fecal from non-fecal streptococci hampered their utilization. Suckling had offered 4 main pieces of evidence in support of fecal streptococci in 1943: [16].

a. Feces of humans and other warm-blooded animals in quite great quantities.

b. Existence in contaminated waters and wastewaters.

c. No presence of environments, pure waters, and pristine soils free of animal and human life.

d. Permanence without reproduction within the environment.

Enumeration of Fecal Streptococci occurred widespread after the useable of Slanetz and Bartley's selective medium in 1957. Thenceforth, many mediums have been suggested to enhance specificity for fecal streptococci and enterococci. Fecal streptococci are characterized by *Streptococcus equinus*, *Streptococcus bovis,* and divers *Enterococcus* spp. The enterococci are the primary indicators of fecal contamination among the fecal streptococci. *Enterococcus faecium*, *Enterococcus faecalis*, *Enterococcus hirae*, and *Enterococcus durans*, are the most common intestinal enterococci. Additionally, certain *Streptococcus* species including *Streptococcus bovis* and *Streptococcus equinus*, and other *Enterococcus* species may be observed on occasion. On the other hand, these streptococci do not live long in water. Because of this enterococci can be used to detect fecal pollution in water [3, 16–19].

Escherichia coli is a thermotolerant coliform that produces indole from tryptophan, however, it is currently also described as coliform that includes the glucuronidase. Furthermore, *E. coli* is identified enzymatically by the absence of urease [17–19].

A wide range of species has existed as fecal contamination indicators. Unfortunately, none have proven to be completely efficient in this function yet. Many, such as *Enterococci*, have a significant environmental occurrence, whilst others, such as *Clostridium perfringens*, are anaerobic through nature or need difficult isolation procedures, such as bacteriophages [16, 17, 19].

4. Detection of bacterial pollution in the water source

Several recent studies have been reviewed culture-dependent methods (heterotrophic plate count, most probable number method, membrane filtration method, and defined substrate/enzyme methods), culture-independent gene sequence-based methods (microbial source tracking, polymerase chain reaction, fluorescence in situ hybridization, and next-generation sequencing, etc.) and immunological methods used to assess bacterial pollution and microbial load in water bodies [17, 21].

4.1 Culture-dependent methods (colorimetric, counting, and fluorimetric analyses)

The most common approaches for the determination of bacterial pollution in environment samples are based on bacterial culturing methods that usually performed biochemical methods required time approximately three days or even longer [22]. The culture-dependent methods applied for the detection of indicator bacteria are schematically shown in **Figure 2**.

The heterotrophic plate count tests (standard plate count/agar plate count) are widely used for estimating the number of live heterotrophic bacteria in various water bodies is a simple procedure that shows the earliest sign of pollution. HPC test methods (pour plate and spread plate methods) are based on incubating at temperature conditions ranging from around 20 to 40°C of agar plates absorbing a small volume

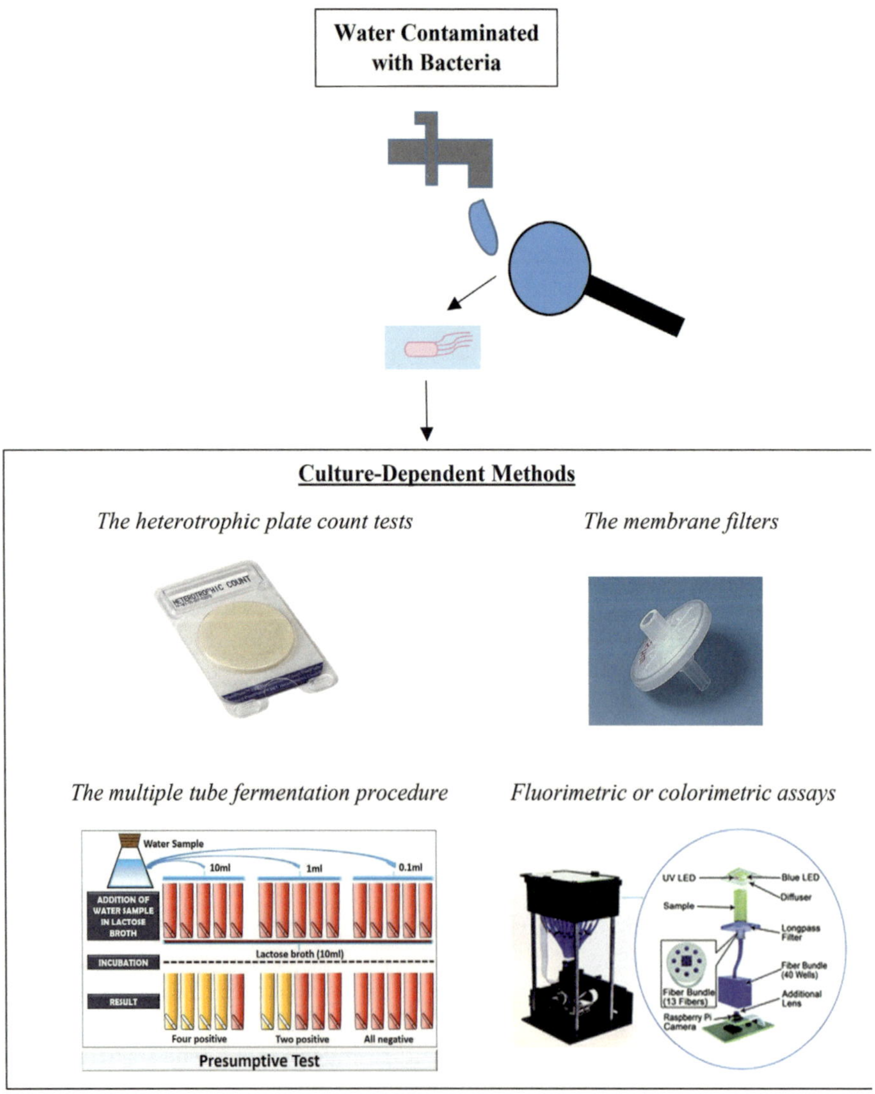

Figure 2.
Some examples of culture-dependent detection methods of indicator bacteria [23–26].

of sample or diluted sample (0.1 to 0.5 mL) All colonies grown on the agar surface and arose from pairs, chains, clusters, or single cells are enumerated and expressed as colony-forming units (CFU) [21, 27].

The multiple tube fermentation procedure is firstly adopted as a bacteriological standard by USA Public Health Service Drinking Water Standard in 1914. This method is indicated the presence of total coliforms, indicators of organic pollution in water sources, and is now referred to as the Most Probable Number (MPN). MPN is illustrated by Prescott et al. [28]. consists of inoculation steps into 15 tubes consisting of 5 tubes for each of three dilution factors (0.1, 1, and 10 mL) each tube containing an inverted Durham tube. This is for detecting total coliforms in the family Enterobacteriaceae relying on the production of acid and gas during lactose fermentation varying with the composition of the media [16, 17]. The coliform pollution in samples is stated as the most probable number (MPN) provided a statistical estimation according to the MPN table [22].

Fluorimetric or colorimetric assays based on specific enzymatic activities are rapid analytical techniques applied for the assessment of total and fecal coliform in water. The enzymatic detection method utilizes the hydrolyzable chromogenic and fluorogenic substrates in the recognition of β-galactosidase in total coliform or β-glucuronidase in *E. coli* [2]. Ortho-nitrophenyl-β-D-galactopyranoside (ONPG) or chlorophenol red-β-D-galactopyranoside (CPRG) are substrates that show a distinct color change in the medium following the β-galactosidase enzyme hydrolysis. A fluorogenic substrate, 4-methylumbelliferyl-β-D-glucuronide (MUG), is hydrolyzed by specific bacterial enzymatic activity to a fluorescence product viewed under ultraviolet (UV) light [29].

The membrane filters in conjunction with Endo-broth for enumerating total coliforms were initiated and used by Mueller in Germany, in 1943 [16]. The water sample is filtered by a membrane with a pore size of 0.45 mm and density may be calculated by counting bacterial cells on the membrane incubated on agar plates. The sensitivity of filtration methods is significantly impressed by the type and quality of membrane filter and the number of colonies.

These culture-based methods that required excessive time (18–96 h) and intensive training for confirmation and verification steps in the detection and quantification of water-origin bacterial pathogens are low sensitivity. Furthermore, the existence of viable but non-culturable (VBNC) bacteria and bacterial pathogens groups at low concentrations in a large volume of water bodies limit efficient quantification and cause false-negative results [3, 12, 30]. Therefore, culture-independent methods having important requirements such as specificity, sensitivity, reproducibility of results, speed, automation, and low cost for reliable analysis are increasingly developing in the last few years [31].

4.2 Immunological methods

Immunological methods are antibody–antigen interactions based on the specific binding affinities of antibodies to specific antigens These methods used polyclonal and monoclonal antibodies include different methods such as lateral flow tests (immunochromatographic assays), SPR (Surface Plasmon Resonance), ELISA (Enzyme-Linked ImmunoSorbent Assay), immunofluorescence, chips, and Western blots etc. (**Figure 3**). The specificity and sensitivity of each technique having the rapid application and specific device requirement depend upon the antibody. However, limitations such as reducing the cell surface antigens, cross-reactivity, false-negative results, and no indication of the viability of organisms for these methods used in the detection of aquatic bacterial pollution are present [12, 21].

4.3 Culture-independent genetic methods

Microbial source tracking (MST) has performed at the end of the 20th century and is a molecular technique to detect the dominant sources of fecal contamination in environmental water samples. Microbial source tracking incorporates unique

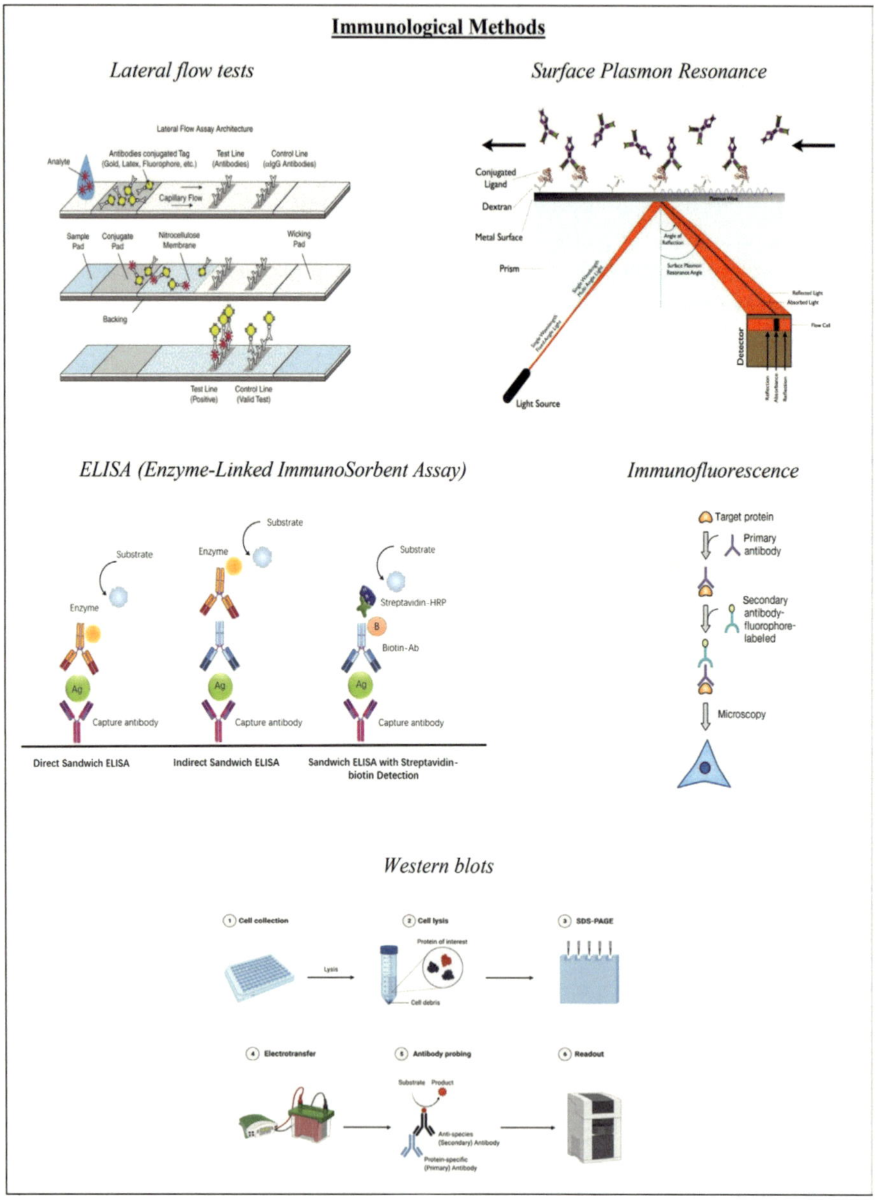

Figure 3.
A view of immunological approaches for bacterial quality of water sources [32–36].

genetic sequences to specific fecal species from fecal sources (e.g. human, dog, cattle) contacted with water bodies The basic approach of MST is the identification of the fecal source using signature molecules (markers) of particular host-associated microorganisms [3, 37]. This method is tested by two basic strategies: 1- library-dependent analyses based on isolation and typing of fecal indicator bacteria for some identifying properties such as their phenotypic (antibiotic resistance, carbon source utilization, etc.) or genotypic fingerprints, and 2- library-independent analyses that required target genes of specific bacterial species such as variable region of the 16S rRNA [37].

Polymerase chain reaction (PCR) has been applied for the detection and identification of total coliform in 100 ml water samples by a screening of the lacZ (developed by Bej et al., wecG, and 16S rRNA genes in the field of water monitoring [2, 38]. Horakova et al. analyzed *β*-d -glucuronidase (*uidA*), lactose permease (*lacY*), *β*-d -galactosidase (*lacZ*), and cytochrome bd complex (*cyd*) four target DNA sequences for specific detection of *E. coli* cells in the water samples. Multiplex PCR, Nested PCR, In situ PCR, and Quantitative real-time PCR (qPCR) methods following the general principle of PCR are used to determine fecal indicator bacteria and microbial load in waters [17, 39]. Detection limits of PCR methods for different water sources are presented in **Table 1**.

Fluorescence in situ hybridization (FISH) reports the presence of fecal contaminant and viable indicator bacteria but nonculturable (VBNC) through hybridization of the sample with rRNA oligonucleotide probes labeled with a fluorescent dye. FISH allows an enumeration of microbial cells in seawater, freshwater, and wastewater samples to obtain qualitative and quantitative results by using fluorescence microscopy, flow cytometry, or confocal microscopy [40].

The most novel approach in the microbial investigation of water quality is **Next-generation sequencing (NGS)** based on the amplification of environmental DNA

PCR detection method	Water-borne bacterial pathogen	Detection limit	Water source
Conventional PCR	*E. coli* Enterotoxigenic *E. coli* (ETEC)	1 cfu/100 mL 4 cfu/mL	Contaminated tap water, Water samples spiked by ETEC and nonpathogenic *E. coli*.
Multiplex PCR	EHEC, *Shigella sp.*, *Vibrio parahaemolyticus*, *P. aeruginosa* and *Salmonella sp.*	10^1 cfu, 10^2 cfu, 10^2 cfu, 10^2 cfu and 10_1 cfu/mL	Polluted water and natural water
Quantitative PCR (qPCR)	*L. monocytogenes*, *V. cholerae*, *V. parahaemolyticus*, *Pseudogulbenkiana* sp., *S. typhimurium*, *S. flexneri*, *C. perfringens* and pathogenic *E. coli*	From 10^2 to 10^4 cells per ca. 200 mg fecal samples of pathogens 100 cells/L	Spiked environmental water samples (pond) and a natural freshwater lake.
Real-time PCR	*V. cholerae*	1 cfu/100 mL	Ballast water.

Table 1.
Limitations of PCR methods for different water sources [12].

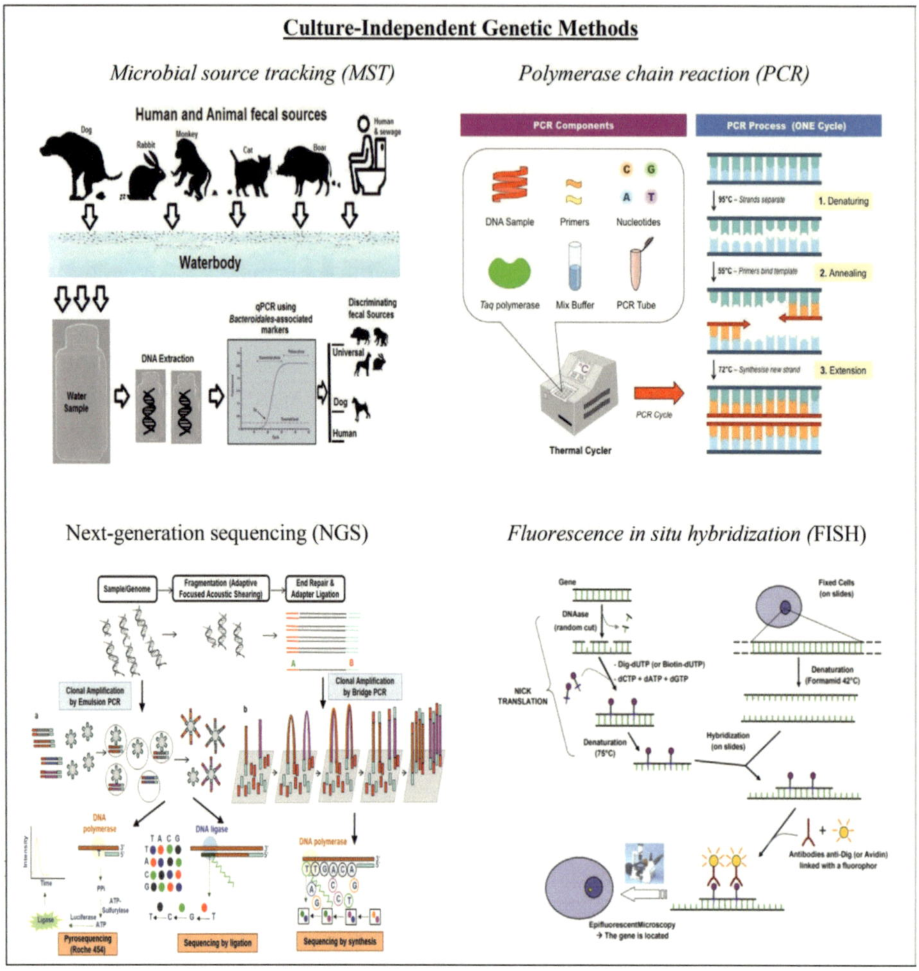

Figure 4.
Molecular strategies for detecting of water quality [41–44].

samples. This method sequenced hypervariable regions (V1, V2, and V4) of unique small subunit SSU rRNA in the microbial communities provides high quality to evaluate fecal pollution [3].

A Diagram of molecular approaches based on genetics is represented in **Figure 4**.

5. Bacterial waterborne diseases

The discharge of waste and stormwater to the surface, coastal, and groundwater, as well as agricultural flows including animal and human waste and nutrients, is raised the microbial pollution of aquatic habitats [30, 45]. Some bacteria among those can ubiquitously occur in many water systems and commonly are recognized waterborne pathogens. Bacterial pathogens including several groups of enteric and aquatic bacteria are classical etiological agents of water-transmitted diseases globally [2, 46]. These pathogens associated with certain infections in humans and animals have been responsible for worldwide tremendous morbidity and mortality [2, 19, 47]. A number

Bacterial agents	Associated diseases	References
Campylobacter spp., *Campylobacter jejuni*	Diarrhea, Gastroenteritis	[12, 19]
Escherichia coli, particularly enterohemorrhagic *E. coli* (EHEC), and others such as enteropathogenic (EPEC), enterotoxigenic (ETEC), and enteroinvasive (EIEC)	Acute diarrhea, bloody diarrhea, and gastroenteritis	[12, 19]
E. coli O157:H7	Bloody diarrhea, hemolytic uremic syndrome	[5]
Legionella pneumophila and related bacteria	Acute respiratory illness, pneumonia (legionellosis)	[12, 19]
Salmonella enterica serotype Typhi	Typhoid fever, paratyphoid, salmonellosis	[19]
Shigella spp.	Bacillary dysentery	[19]
Vibrio cholerae	Cholera, Gastroenteritis	[12, 19]
Yersinia enterocolitica	Gastroenteritis/Yersiniosis	[1, 19]
Helicobacter pylori	Abdominal pain, peptic ulcers, gastric cancer, and chronic gastritis,	[5, 12]
Burkholderia pseudomallei	Meliodosis	[12]
Non-tuberculous mycobacteria	Pulmonary disease, skin infection	[12]
Pseudomonas aeruginosa	Infections on the lungs, urinary tract, and kidney, inflammation, and sepsis	[12]

Table 2.
Some potential examples of the bacterial water-transmitted diseases.

of reasons such as the excessive increase in population, globalization of travel and commerce as well as contamination of drinking water are caused the appearance again and again of water-associated diseases. Especially, the lack of technological and financial resources in developing countries contributes to emerging waterborne outbreaks [12]. Children under 5 years of age are particularly influenced by diarrheal infections transmitted by contaminated drinking water in African and Asian countries. It is also referred to has no access to safe and clean drinking water and dying more than 1.5 million children per year from contaminated drinking water [19].

Vibrio cholerae and *Salmonella enterica* serovar *Typhi* (now known as *Salmonella typhi*) were identified as the first water-borne pathogens in the 19 th century [2]. A list of declared bacterial water-transmitted diseases from the 19th century to the present is shown in **Table 2**.

6. Conclusions

The aquatic environments are important ecosystems threatened by harmful pollutants (metals, agrochemicals, nanoparticles, radioactive elements, volatile organic compounds, personal care products, household products, industrial solvents, and waterborne pathogens) due to anthropogenic residues and lack of sanitation. For this reason, the regular control, monitoring, understanding, and evaluation of water quality is a global concern for environmental and public health safety. One of the most parameters of water quality in terms of health risk assessment is, indicator bacteria are key to determining microbiological perspective. Culture-dependent methods, molecular methods, and immunological techniques for estimating bacterial load and populations in water systems, culture-dependent are routinely utilized. Despite having several disadvantages (costly, excessive time consuming, the lack of protocol and sample processing standardization, etc.) of each method, molecular techniques are the most effective strategies to identify causative agents to water-related diseases and waterborne outbreaks and the distribution of proxy bacteria in water bodies.

Author details

Hatice Aysun Mercimek Takci[1], Melis Sumengen Ozdenefe[2], Fikret Buyukkaya Kayis[3] and Sadık Dincer[4*]

1 Kilis 7 Aralık University, Kilis, Turkey

2 Near East University, Nicosia, Northern Cyprus

3 Adiyaman University, Adiyaman, Turkey

4 Cukurova University, Adana, Turkey

*Address all correspondence to: sdincer@cu.edu.tr

References

[1] Lugo JL, Lugo ER, Puente MDL. A systematic review of microorganisms as indicators of recreational water quality in natural and drinking water systems. Journal of Water and Health. 2021;**19**(1):20-28. DOI: 10.2166/wh.2020.179

[2] Some S, Mondal R, Mitra D, Jain D, Verma D, Das S. Microbial pollution of water with special reference to coliform bacteria and their nexus with environment. Energy Nexus. 2021;**1**:100008. DOI: 10.1016/j.nexus.2021.100008

[3] Motlagh AM, Yang Z. Detection and occurrence of indicator organisms and pathogens. Water Environment Research. 2019;**91**(10):1402-1408. DOI: 10.1002/wer.1238

[4] Resend JA, Lúcia da Silva V, Diniz CG. Aquatic environments in the one health context: Modulating the antimicrobial resistance phenomenon. Acta Limnologica Brasiliensia. 2020;**32**:e102. DOI: 10.1590/S2179-975X4719

[5] Bashir I, Lone FA, Bhat RA, Mir SA, Dar ZA, Dar SA. Concerns and threats of contamination on aquatic ecosystems. In: Bioremediation and Biotechnology. Cham: Springer; 2020. pp. 1-26. DOI: 10.1007/978-3-030-35691-0_1

[6] Matyar F, Gülnaz O, Guzeldag G, Mercimek H, Akturk S, Arkut A, et al. Antibiotic and heavy metal resistance in gram-negative bacteria isolated from the Seyhan Dam Lake and Seyhan River in Turkey. Annals of Microbiology. 2014;**64**(3):1033-1040. DOI: 10.1007/s13213-013-0740-8

[7] Hampel M, Blasco J, Segner H. Molecular and cellular effects of contamination in aquatic ecosystems. Environmental Science and Pollution Research. 2015;**22**(22):17261-17266. DOI: 10.1007/s11356-015-5565-5

[8] Omer NH. Water quality parameters. In: Water quality-science, assessments and policy. Intechopen. 2019;**2019**:1-18. DOI: 10.5772/intechopen.89657

[9] Sadler MC, Senouillet J, Kuenzi S, Grasso L, Watson DC. Computational surveillance of microbial water quality with online flow cytometry. Frontiers in Water. 2020;**2**:1-12. DOI: 10.3389/frwa.2020.586969

[10] Singh RK, Iqbal SA, Seth PC. Bacteriological pollution in a stretch of river Narmada at Hoshangabad, Madhya Pradesh. Pollution Research. 2001;**20**(2):211-213

[11] Mercimek Takci HA, Toplar S, Sumengen OM. Antibiotic and heavy metal resistance of *Escherichia coli* strains isolated from the Seve Dam and Konak Pond, Kilis, Turkey. Acta Aquatica Turcica. 2021;**17**(2):290-297. DOI: 10.22392/actaquatr.801564

[12] Ramírez-Castillo FY, Loera-Muro A, Jacques M, Garneau P, Avelar-González FJ, Harel J, et al. Waterborne pathogens: Detection methods and challenges. Pathogens. 2015;**4**(2):307-334. DOI: 10.3390/pathogens4020307

[13] Namoi Catchment Management Authority. Water Quality Parameters and Indicator. New South Wales: Stephanie McCaffrey; 2013 Available from: http://www.waterwatch.nsw.gov.au

[14] Shah CR. Which physical, chemical and biological parameters of water determine its quality?

[Internet]. 2017. Available from: https://www.researchgate.net/publication/317588226_Which_Physical_Chemical_and_Biological_Parameters_of_water_determine_its_quality/stats [Accessed June 14, 2017]. 10.13140/RG.2.2.29178.90569

[15] Shaltami OR, Bustany I. Water quality – A review. In: Proceedings of the 2nd International Symposium on Geosciences (ISG2021). Brazil: Ponta Grossa; 2021. pp. 56-62

[16] Ashbolt NJ, Grabow WO, Snozzi M. Indicators of microbial water quality. In: Fewtrell L, Bartram J, editors. Water Quality: Guidelines, Standards and Health: Assessment of Risk and Risk Management for Water-related Infectious Disease. London: Published by IWA Publishing; 2001. pp. 289-316

[17] Tallon P, Magajna B, Lofranco C, Leung KT. Microbial indicators of faecal contamination in water: A current perspective. Water, Air, and Soil Pollution. 2005;**166**(1):139-166. DOI: 10.1007/s11270-005-7905-4

[18] Figueras M, Borrego JJ. New perspectives in monitoring drinking water microbial quality. International Journal of Environmental Research and Public Health. 2010;**7**(12):4179-4202. DOI: 10.3390/ijerph7124179

[19] Saxena G, Bharagava RN, Kaithwas G, Raj A. Microbial indicators, pathogens and methods for their monitoring in water environment. Journal of Water and Health. 2015;**13**(2):319-339. DOI: 10.2166/wh.2014.275

[20] Ohrel RL, Register KM. Bacteria: Indicators of potential pathogens. In: Volunteer Estuary Monitoring Manual: A Methods Manual. Second ed. Washington: Office of Wetlands, Oceans and Watersheds to The Ocean Conservancy; 2006. pp. 1-20

[21] Tanchou V. Review of methods for the rapid identification of pathogens in water samples: ERNCIP thematic area Chemical & Biological Risks in the Water Sector Task 7, deliverable 1. Luxembourg: Publications Office of the European Union. [Internet]. Joint Research Centre, Institute for the Protection and Security of the Citizen; 2014. Available from: DOI: 10.2788/18775 [Accessed: February 20, 2015]

[22] Canciu A, Tertis M, Hosu O, Cernat A, Cristea C, Graur F. Modern analytical techniques for detection of bacteria in surface and wastewaters. Sustainability. 2021;**13**(13):7229. DOI: 10.3390/su13137229

[23] Treatment Plant Operator. The Heterotrophic Plate Count Test [Internet]. 2017. Available from: https://www.tpomag.com/online_exclusives/2017/11/the_heterotrophic_plate_count_test_sc_001fa [Accessed November 14, 2017]

[24] Analytics shop.com. Membrane filters, 0.2 μm, sterile [Internet]. Available from: https://www.analytics-shop.com/gb/br26530-gb.html

[25] Biology Reader. Most Probable Number Method [Internet]. 2022. Available from: https://biologyreader.com/most-probable-number-method.html

[26] Tok S, De Haan K, Tseng D, Usanmaz CF, Koydemir HC, Ozcan A. Early detection of *E. coli* and total coliform using an automated, colorimetric and fluorometric fiber optics-based device. Lab on a Chip. 2019;**19**:2925-2935. DOI: 10.1039/C9LC00652D

[27] 9215 Heterotrophic Plate Count, Standard Methods for the Examination of Water and Wastewater. Available from: https://www.standardmethods.org/doi/full/10.2105/SMWW.2882.188 [Accessed February 15, 2021]

[28] Prescott LM, Harley JP, Donald AK. Microbiology. 5th ed. New York, NY, USA: W.C. Brown Publishers, McGraw-Hill Companies Inc.; 1996

[29] Rodrigues C, Cunha MÂ. Assessment of the microbiological quality of recreational waters: Indicators and methods. Euro-Mediterranean Journal for Environmental Integration. 2017;**2**(1):1-18. DOI: 10.1007/s41207-017-0035-8

[30] Fitzmorris-Brisolara K, Maal-Bared R, Worley-Morse T, Danley-Thomson A, Sobsey M. Monitoring coliphages to reduce waterborne infectious disease transmission in the one water framework. International Journal of Hygiene and Environmental Health. 2022;**240**:113921. DOI: 10.1016/j.ijheh.2022.113921

[31] Kostić T, Stessl B, Wagner M, Sessitsch A. Microarray analysis reveals the actual specificity of enrichment media used for food safety assessment. Journal of Food Protection. 2011;**74**(6):1030-1034. DOI: 10.4315/0362-028X.JFP-10-388

[32] Cytodiagnostics. Lateral Flow Assay [Internet]. 2022. Available from: https://www.cytodiagnostics.com/pages/lateral-flow-assay

[33] Wikipedia. Surface plasmon resonance [Internet]. 2022. Available from: https://en.wikipedia.org/wiki/Surface_plasmon_resonance [Accessed May 24, 2022]

[34] Creative Biolabs. Sandwich ELISA with Direct Detection [Internet]. 2022. Available from: https://www.antibody-creativebiolabs.com/sandwich-elisa-with-direct-detection.htm

[35] Ryu WS. Diagnosis and methods. In: Molecular Virology of Human Pathogenic Viruses. Elsevier, Academic Press; 2017. pp. 47-62. DOI: 10.1016/B978-0-12-800838-6.00004-7

[36] Rockland. Western Blotting Technique [Internet]. 2021. Available from: https://www.rockland.com/resources/western-blotting-technique/

[37] Harwood VJ, Staley C, Badgley BD, Borges K, Korajkic A. Microbial source tracking markers for detection of fecal contamination in environmental waters: Relationships between pathogens and human health outcomes. FEMS Microbiology Reviews. 2014;**38**(1):1-40. DOI: 10.1111/1574-6976.12031

[38] Bej AK, Dicesare JL, Haff L, Atlas RM. Detection of *Escherichia coli* and *Shigella* spp. in water by using the polymerase chain reaction and gene probes for uid. Applied and Environmental Microbiology. 1991;**57**(4):1013-1017. DOI: 10.1128/aem.57.4.1013-1017.1991

[39] Horakova K, Mlejnkova H, Mlejnek P. Specific detection of *Escherichia coli* isolated from water samples using polymerase chain reaction targeting four genes: cy- tochrome bd complex, lactose permease, *β*-d-glucuronidase, and *β*-d-galactosidase. Journal of Applied Microbiology. 2008;**105**(4):970-976. DOI: 10.1111/j.1365-2672.2008.03838.x

[40] Kuo JT, Chang LL, Yen CY, Tsai TH, Chang YC, Huang YT, et al. Development of fluorescence in situ hybridization as a rapid, accurate method for detecting coliforms in water samples. Biosensors. 2020;**11**(1):8. DOI: 10.3390/bios11010008

[41] Swuertzlab. Microbial source tracking, pathogen detection, QMRA [Internet]. Available from: https://swuertzlab.com/microbial-source-tracking-pathogen-detection-qmra/

[42] Bioninja. Overview of a Polymerase Chain Reaction Cycle [Internet]. 2021. Available from: https://ib.bioninja.com.au/standard-level/topic-3-genetics/35-genetic-modification-and/pcr.html

[43] Gupta N, Verma VK. Next-generation sequencing and its application: Empowering in public health beyond reality. In: Arora P, editor. Microbial Technology for the Welfare of Society. Microorganisms for Sustainability. Vol. 17. Singapore: Springer; 2019. pp. 313-341. DOI: 10.1007/978-981-13-8844-6_15

[44] Flogentec. Fluorescence in situ hybridization (FISH) The Interne 2019. Available from: https://www.flogentec.com/en/biology-experiment-fields-of-application/fluorescence-in-situ-hybridization/

[45] Buyukkaya KF, Dincer S, Matyar F, Mercimek Takci HA, Sumengen OM. An investigation of multiple antibiotic resistance and identification of Bacteria isolated from Gölbaşı and Azaplı Lakes (Adıyaman). Turkish Journal of Agriculture-Food Science and Technology. 2017;**5**(1):43-47. DOI: 10.24925/turjaf.v5i1.43-47.819

[46] Magana-Arachchi DN, Wanigatunge RP. Ubiquitous waterborne pathogens. In: Waterborne Pathogens. Elsevier, Butterworth-Heinemann; 2020. pp. 15-42. DOI: 10.1016/B978-0-12-818783-8.00002-5

[47] Buyukkaya Kayis F, Dincer S, Matyar F. Isolation, identification and antibiotic resistance of *Aeromonas* spp. and *Pseudomonas aeruginosa* from Azapli and Golbasi Lakes, Adiyaman, Turkey (Southeast Anatolian region). Fresenius Environmental Bulletin. 2018;**27**(1):116-124

Chapter 7

Chemical Quality of Rainwater and Surface Runoff Water in the Proximity of the Abstractions That Supply the São Pedro do Sul Medical Spa (Portugal)

L.M. Ferreira-Gomes, Ana I.S. Tavares Jorge,
Liliana F. Gonçalves and S.M. Soares Almeida

Abstract

São Pedro do Sul medical spa provides health services using hot natural groundwater superiorly classified as natural mineral water. The main source of supply of that medical spa, has been over time, the Traditional Spring, which in recent decades has been systematically controlled, and shows spectacular constancy in its chemical quality, if there are no outside contaminations. Within the context presented, the detailed physical-chemical quality of rainwater and surface runoff water was studied, in the proximity of the Traditional Spring, as there is a potential for them to infiltrate at depth and evolve into the natural mineral water aquifer system. Thus, in the present chapter, after presenting the physical-chemical quality of the natural mineral water from the Traditional Spring, as well as some elements of the literature on the physical-chemical quality of rainwater, the methodology of work is followed, and then the results obtained from the physical-chemical composition of rainwater and surface run-off water are presented and discussed, comparing them with the quality of the water from the Traditional Spring. Finally, the main conclusions are presented, and some recommendations are made on research into potential sources of pollution, which justify the poor quality of the rainwater studied.

Keywords: chemical composition, rainwater, surface runoff water, natural mineral water, São Pedro do Sul medical spa

1. Introduction

The medical spa of São Pedro do Sul (SPS), in Portugal, with a millenary tradition, was already used by the Romans about 2000 years ago. Currently, it is a great asset to the region, as it allows leveraging the local economy, due to its activity in thermalism and other associated and complementary activities, such as tourism. Thermalism is the activity practiced in the medical spa, namely the use of natural mineral water (NMW) in some treatments, in the field of balneotherapy. SPS medical spa has two operating bathhouses (Thermal Centre and Queen D. Amélia Bathhouse) and has already reached around 25,000 users in classic thermalism per year; each user of classical thermal treatments attends a Bathhouse for about 2 weeks. In this process, the direct income with the treatments has already reached about 5 million euros/year [1].

The main source to supply the SPS medical spa, is the Traditional Spring (NT), which provides special groundwater with about 10 L/s of the sulphureous water type. There is also a well (AC1) that complements the supply of groundwater similar to the NT water. It is important to note that the water in the NT comes from a semi-confined aquifer system, of granite rocks, of the fissure type, very deep; its water, initially rain, infiltrates to great depths, in the order of several kilometers, and on its way acquires its specific chemical quality, function of the water/rock interaction, also acquiring temperatures above 100°C; in that process, and due to the geological singularities, the NMW resurfaces in the NT at about 68°C.

There is also, the superficial aquifer system, unconfined, in granitic rocks, more common in the region, with about 100 m depth, essentially of fissure type, and sometimes porous, when the granites are very weathered; in this superficial aquifer system, its waters are slightly acidic (pH <6.0) and generally have total mineralization (M_T) lower than 250 mg/L. Detailed studies on these situations can be seen in works by several authors [2, 3]. The chemical composition of the special groundwater, from NT, is very well characterized and is classified by the Portuguese state as a NMW for thermalism applications. The guarantee of its physical-chemical stability, besides other requirements, allows the use of that water as an equivalent of a medicine, with treatments for rheumatic and respiratory diseases being prescribed by a doctor.

The chemical composition of that water is analyzed over time and in a systematic way. Such quality control allows you to check, if anomalies are occurring, being sometimes its enigmatic causes.

Records in 2005, from physical-chemical analyses of the water of the NT, identified values of its trace component out of the common, namely of lead (Pb), zinc (Zn) and copper (Cu). This situation led to several interventions, namely the study of the chemical quality of the rainwater and surface runoff water in the area surrounding the NT. Thus, it is emphasized that it is the chemical quality of rainwater and surface runoff water in the medical spa area that this chapter focuses on, in order to try to understand whether changes in the water quality of the NT may result from them.

The SPS medical spa is located in an area called "Polo das Termas" of the São Pedro do Sul Hydromineral and Geothermal Field—CHGSPS (**Figure 1**).

The territory's characteristics are essentially a dispersed urban space, with some equipment for medical spa activities, and other associated and consequent ones, such as hotels, restaurants, and tourism. Green spaces predominate, although most of them present characteristics of brush and poorly regulated forest, interspersed with small family agricultural areas or scattered patches of houses.

The nearest town is São Pedro do Sul, about 3.5 km NE of the medical spa area. The town of São Pedro do Sul has only about 3600 inhabitants, being the municipality seat, with 14 parishes, in a global territory with 348.95km^2 of the area and 16,851 inhabitants (2011). The municipality of São Pedro do Sul, as well as the neighboring ones, each has 1 to 2 industrial parks with various factories, car workshops, and services, but without any apparent expression capable of causing record air pollution. There is also no known focus on air pollution in those municipalities.

About the chemical quality of the natural mineral water of the NT, it is presented in **Table 1** statistics of the results of physical-chemical analysis, between 1985 and

Figure 1.
Geographic framing of the study area (a) and image of SPS medical spa zone, with the location of its abstractions (Main spring—NT, and well AC1), and of its bathhouses (thermal center and queen D. Amélia bathhouse (b).

2014, including the results, of the most anomalous phase (2005–2006), and also of several years following the problem, which is now resolved.

The parameters that present a SD_R lower than 10% correspond to relatively very stable chemical elements. One obvious case in the opposite situation is Aluminium (Al), which presents an SD_R of 217%. This happens because although the water naturally includes some Al content, it sometimes appears with abnormal values, being understood as a consequence of local anthropic actions; in 2005–2006, and following years, there was no problem with Al. Other cases, clearly evident in bad situations, are those in which the values of certain chemical elements are suddenly very high, when the normal content of the chemical element being studied is below the detection limit, as is the case with Pb, Zn, and Cu.

That context led to the creation of an External Monitoring System (EMS) in the area surrounding the NT, which was presented in detail in: "Good practices of quality control in the area surrounding of natural mineral water abstraction of São Pedro do Sul medical spa (Portugal)" [1], and which will be presented in the methodology section of this chapter.

On hydrological elements of the study area, based on records between 1933 and 1960, in the São Pedro do Sul udometric station, the average annual precipitation was 1103 mm. The average annual air temperature in that region is around 13°C. From the monthly sequential hydrological balance for the same region, it is obtained annual surpluses of 675 mm [5]. In **Figure 2**, the evolution of precipitation throughout the year is presented, being noteworthy that the month with the lowest precipitation is July, with 11.8 mm, and the month with the highest precipitation is March with 175 mm.

Rainwater and surface runoff water quality studies in the context of the present work, as far as the authors know, are inexistent, because medical spas are normally located in special equipment areas associated with tourism and leisure, and in green parks, so there is no concern that rainwater in these areas may have harmful chemical elements; the fact is that if this happens, the water infiltrating around the natural mineral water abstractions areas may contaminate the resource and create public health problems. Studies on rainwater quality in contexts different from the present chapter are frequent in the literature; some considerations about this domain are presented in the follow-up, to facilitate the framing of the results of this study. Rainwater contains small amounts of dissolved and suspended substances. The

Parameter		N	min	ave	max	SD	SD_R (%)
pH		96	8.33	8.82	8.95	0.08	1
Conductivity – C (μS/cm)		96	351.00	405.08	485.00	26.14	6
Total Sulphuration (in I_2 0.01 N-mL/L)		96	16.00	21.30	34.00	3.70	19
Total Alkalinity—A_T (in HCl 0.1 N-mL/L)		94	22.00	23.68	25.00	0.70	3
Total Hardness-H_T (in p.p.10^5 of $CaCO_3$)		96	0.65	0.77	1.10	0.06	8
Total CO_2 (mmol/L)		80	1.81	2.06	2.50	0.09	4
Silica—SiO_2 (mg/L)		96	60.90	67.94	78.50	3.61	5
Dry Residue – R_D (at 180°C) – (mg/L)		95	291.30	304.77	326.00	6.27	2
Total Mineralization – M_T (mg /L)		96	333.00	359.68	385.00	8.68	2
Cations (mg/L)	Sodium (Na^+)	96	85.40	90.19	96.00	1.93	2
	Calcium (Ca^{2+})	96	1.60	2.99	4.40	0.30	10
	Potassium (K^+)	93	2.90	3.30	3.70	0.16	5
	Magnesium (Mg^{2+})	55	< 0.03	—	< 1.0	—	—
	Lithium (Li^+)	93	0.47	0.59	0.70	0.04	7
	Ammonium (NH_4^+)	96	0.16	0.33	0.47	0.04	13
	Iron (Fe^{2+})	21	< 0.075	—	0,100	—	—
Anions (mg/L)	Bicarbonate (HCO_3^-)	96	101.90	119.48	133.00	5.48	5
	Chloride (Cl^-)	96	25.10	27.89	37.00	1.53	5
	Sulphate (SO_4^{2-})	94	7.80	10.12	13.00	0.95	9
	Fluoride (F^-)	96	15.20	17.71	19.00	0.56	3
	Carbonate (CO_3^{2-})	96	3.00	4.69	9.50	1.10	23
	Nitrate (NO_3^-)	95	< 0.05	—	0.97	—	—
	Nitrite (NO_2^-)	96	< 0.002	—	< 0.02	—	—
	Bisulfide (HS^-)	96	1.70	3.26	5.60	0.64	20
	Silicate ($H_3SiO_4^-$)	87	7.00	11.89	15.10	1.69	14
	Phosphates ($H_2PO_4^-$)	3	< 0.04	—	< 0.04	—	—
Trace elements (mg/L)	Silver (Ag)	49	< 0.00004	—	< 0.0005	—	—
	Aluminium (Al)	51	0.0021	0.0437	0.6400	0.0950	217
	Arsenic (As)	53	0.0030	0.0045	0,0180	0.0022	49
	Boron (B)	51	0.3600	0.4263	0.4760	0.0245	6
	Barium (Ba)	53	< 0.0003	—	0.0063	—	—
	Beryllium (Be)	53	0.0003	0.0006	0.0014	0.0002	28
	Bismuth (Bi)	40	< 0.00002	—	0.0001	—	—
	Bromide (Br)	3	0.1300	0.1750	0.2600	0.0601	34
	Cadmium (Cd)	53	< 0.00002	—	0.0006	—	—

Parameter	N	min	ave	max	SD	SD_R (%)
Cobalt (Co)	53	< 0.00001	—	0.0003	—	—
Chromium (Cr)	53	< 0.0004	—	0.0124	—	—
Cesium (Cs)	42	0.0460	0.0624	0.0721	0.0043	7
Copper (Cu)	51	< 0.00005	—	0.0550	—	—
Mercury (Hg)	47	< 0.00007	—	0.0003	—	—
Iodide (I)	3	< 0.00006	—	0.0020	—	—
Manganese (Mn)	52	0.0013	0.0020	0.0060	0.0009	45
Molybdenum (Mo)	53	< 0.001	—	0.0060	—	—
Niobium (Nb)	52	<0.00002	—	0.0001	—	—
Nickel (Ni)	52	< 0.0002	—	0.0150	—	—
Lead (Pb)	52	< 0.00006	—	0.0730	—	—
Rubidium (Rb)	42	0.0530	0.0590	0.0670	0.0028	5
Antimony (Sb)	53	< 0.00009	—	0.0070	—	—
Selenium (Se)	49	< 0.00085	—	0.0013	—	—
Tin (Sn)	53	< 0.00003	—	0.0010	—	—
Strontium (Sr)	50	0.0570	0.0667	0.0730	0.0032	5
Tantalum (Ta)	41	< 0.00001	—	0.00004	—	—
Tellurium (Te)	48	< 0.00005	—	0.0003	—	—
Thallium (Tl)	41	< 0.0001	—	0.0015	—	—
Uranium (U)	47	< 0.00002	—	0.0001	—	—
Vanadium (V)	53	< 0.0002	—	0.0009	—	—
Yttrium (Y)	53	< 0.00001	—	0.00002	—	—
Tungsten (W)	53	0.0370	0.0791	0.0990	0.0101	13
Zinc (Zn)	53	< 0.0002	—	0.3140	—	—
Zirconium (Zr)	41	< 0.00015	—	0.0002	—	—

N: number of samples; min: minimum, ave.: average, max: maximum, SD: standard deviation, and SD_R: relative standard deviation (SD/ave) × 100.

Table 1.
Statistics of the results of physical-chemical analyses of the natural mineral water collected in the traditional spring (NT), between 1985 and 2014 [4].

concentration of these substances as well as the pH can vary with time [7] and place [8]. Another very common situation in rainwater quality studies is related to the content of some chemical elements as a function of distance from the sea, namely Cl^- and Na^+. The most important compounds that can be found in rainwater are Cl^-, SO_4^{2-}, NO_3^-, NO_2^-, and HCO_3^- in anions, and Na^+, K^+, Ca^{2+}, Mg^{2+} and NH_4^+ in the cations [9]. Nowadays, the existence of immense industrial parks, of different types, spread all over the world, as well as the burning of fossil fuels for energy production, locomotion of vehicles and airplanes, or even the incineration of dangerous waste and others, leads to enrichment in the atmosphere of very potentially harmful chemical elements, which without these anthropogenic sources, they would never appear.

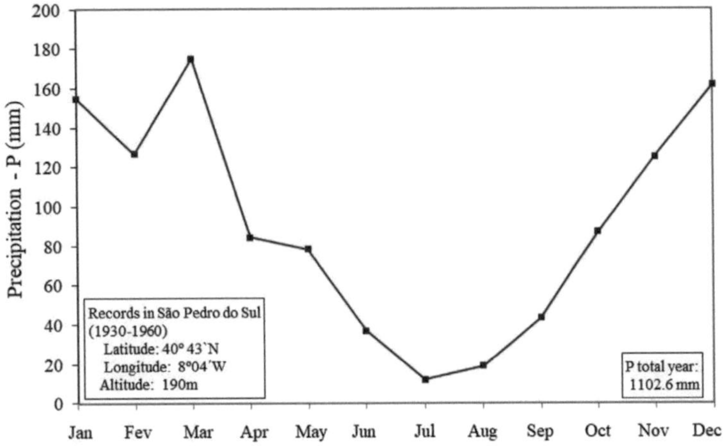

Figure 2.
Evolution of the average monthly precipitation in the region of SPS (from [6]).

There have been many studies on the quality of rainwater in different situations under the action of anthropic factors. The works of several authors dealing with the quality of rainwater in different situations are worth mentioning, showing very problematic cases [10–12], especially in situations where there is a need to collect rainwater to supply populations. In synthesis, the purity of rainwater in any region of the world should not be guaranteed whenever public health situations are involved.

2. Methodology

In this item it is important to mention that the natural mineral water abstractions (Traditional Springer—NT and Well AC1) are located in a garden zone, immediately upstream of the Queen D. Amelia Bathhouse (**Figure 3**). Surrounding that garden area, and upstream of it, there is a public street (Thermas garden street) that includes parking capacity for about 30 cars, besides the location of a hotel (Hotel do Parque), contiguously, with outdoor parking. The area surrounding the NT house, and in particular in the upstream sector (area NW and SW of the house) at the time when there were problems in the quality of the NT natural mineral water, had luxuriant vegetation and in particular banana trees. Following the detection of problems in the quality of NT natural mineral water, interventions were planned, organized into two groups:

2.1 Immediate interventions

I. to start using the NT as the main abstraction, because since 2002, with the beginning of the exploration of the young well AC1, the NT was only a complement; it should be noted that the natural mineral water aquifer system is of the semi-confined type, and the exploration of one abstraction interferes with the other; thus, from 2005 onwards, a great constraint was imposed on the exploitation of Well AC1, sometimes closing it totally, in order to raise the piezometric level of the natural mineral water aquifer system; in this way the infiltrations of rainwater and other waters in the garden area were minimized;

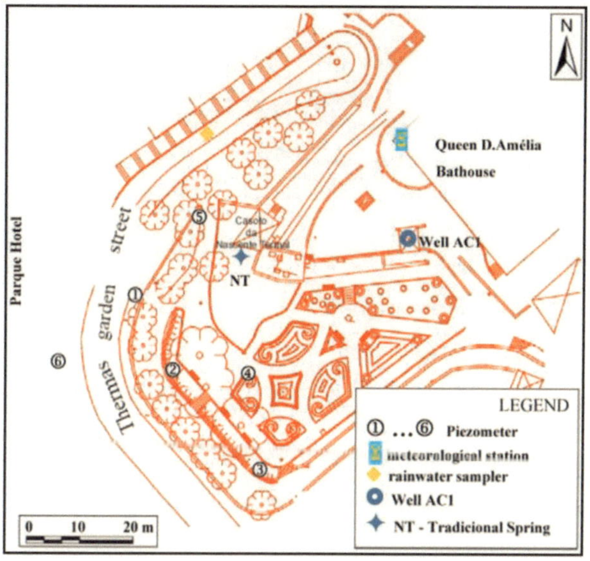

Figure 3.
Location of external monitoring equipment in the upstream zone of the natural mineral water abstractions of the SPS medical spa.

II. placing a gutter on the roof of the NT access house on the south and west sides, and harmonizing the situation with the surface runoff water so that it does not infiltrate locally; the external area of the roof of the NT house was also waterproofed;

III. check the drainage of surface groundwater in the surroundings of the NT house, in order to ensure that there are no leaks or infiltrations into the NT.

The interventions in (III) had unforeseen developments, very prolonged in time, and in particular because many meticulous excavations were carried out around the NT house, for having found underground water resurgences equal to that of the NT, which immediately interfered with its flow; part of the garden was eliminated, mainly the area with banana trees, substituting very organic earth, and others, namely by adequate clays that waterproofed the surrounding of the NT house.

2.2 Short-term intervention

Install an External Monitoring System (EMS), to control what happens over time to the environmental and groundwater quality resulting from possible infiltration from various sources such as agriculture or gardening, vehicle traffic, storm-water networks, or even the wastewater network.

Synthetically, the EMS consists of three main components: (i) piezometers for monitoring groundwater levels, conductivity, temperature, and for collecting water for physical-chemical quality control, namely heavy metals; (ii) a surface runoff water sampling system for physical-chemical analysis of the water resulting from the runoff of the street area; and (iii) a meteorological station with a precipitation collection system for physical-chemical analysis.

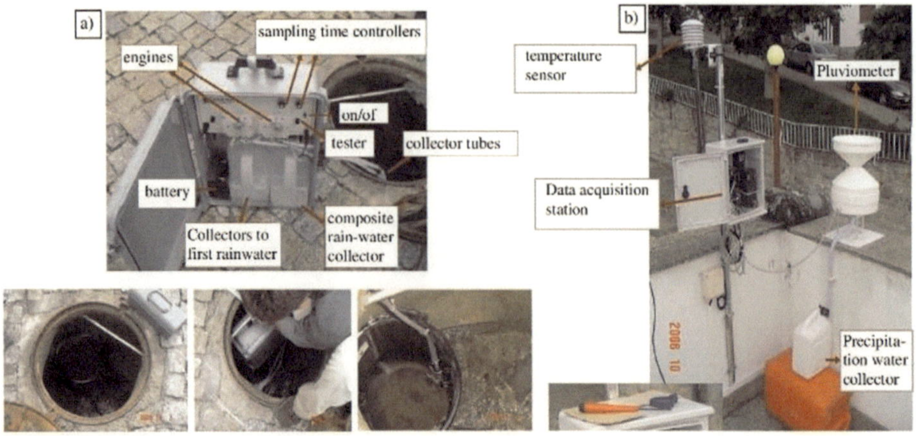

Figure 4.
Images of monitoring systems: (a) street surface runoff water sampling system; and (b) meteorological station with a precipitation collector.

The surface runoff water sampling system, installed in the Thermas garden street (**Figure 4**), consists of a PVC tank for storing surface runoff water, to which an automatic sampler is associated for taking water samples. The automatic sampler has two independent sampling circuits, allowing the differentiated collection in two polyethylene containers, each with a capacity of four liters. The two samplers made it possible to organize the samples with two designations: the simple sample, which is the one resulting from the first rainfall that goes into the PVC tank and from there is sucked continuously into the sampler; the composite sample results from intermittent sampling over time. Regarding the meteorological station, it was installed on the terrace edge of the Queen D. Amélia Bathhouse, in a place with the following coordinates: Latitude $-40°44'20''$N, Longitude $-8°05'34''$W, Altitude: 154 m. This point is about 55 km east of the coastline (Atlantic Ocean). That station consists of a modular data acquisition system which, among others, has a tilting type rainfall sensor with a sensitivity of 0.2 mm per tilting. The coating is made of UV-resistant plastic, so as not to alter the chemical parameters of the water and rain and is associated with a container to retain the water, thus enabling its subsequent analysis in the laboratory.

In order to evaluate the quality of rainwater and surface runoff water, it was established as a minimum sampling period of 1 year, with monthly collections, however, the period ended up being higher, because it did not always rain on the date scheduled for the collections, in addition to some occasional constraints of technical and logistical nature, so in order to have greater representativity of the results, the sampling period ended up being between October 2006 and October 2008. It should be mentioned that in this period systematic analyses of groundwater from the piezometers were also carried out, however, these are not the subject of studies in this chapter.

The physical-chemical analyzes were carried out almost globally by the Laboratory of INETI (current Laboratory of LNEG), except for the analyzes of the first two campaigns, which were carried out by the Laboratory of IST, still in 2006. Those laboratories are accredited in accordance with the rules of the European Union. In summary, the parameters to be researched were organized as follows: (i) global parameters (pH, conductivity-C, total alkalinity-A_T, total hardness-H_T, Silica-SiO_2, and total mineralization-M_T); (ii) major cations (Sodium-Na^+, Calcium-Ca^{2+},-Potassium-K^+, Magnesium-Mg^{2+}, Lithium-Li^+, Ammonium-NH_4^+, Iron-Fe^{2+}); (iii) major anions (Bicarbonate-HCO_3^-, Chloride-Cl^-, Sulphate-SO_4^{2-}, Fluoride-F^-, Nitrate-NO_3^-, Nitrite-NO_2^-, Phosphates-$H_2PO_4^-$); and (iv) trace elements, such as details, for example, are presented in **Figure 5**. The analytical methods followed were:

Parameter		N	Min	Ave	Max	SD	SD_R (%)
pH		12	4.17	5.27	5.93	0.47	9
Conductivity - C (μS/cm)		12	8.00	32.03	75.00	22.46	70
Total Alkalinity - A_T (in HCl 0.1N - mL/L)		12	< 0,05	-	1.00	-	-
Total Hardness - H_T (in p.p.10^5 $CaCO_3$)		10	0.14	1.13	4.30	1.28	113
Sílica - SiO_2 (mg/L)		12	< 0.3	-	1.10	-	-
Total Mineralization - M_T (mg /L)		9	2.20	9.91	24.40	6.39	64
C	Sodium (Na^+)	12	0.20	1.65	4.90	1.44	88
a	Calcium (Ca^{2+})	12	0.14	1.78	7.40	2.05	116
t	Potassium (K^+)	11	0.06	0.26	1.20	0.31	122
i	Magnesium (Mg^{2+})	12	0.04	0.31	0.90	0.28	89
o	Lithium (Li^+)	10	0.0001	0.0011	0.0033	0.0010	92
n	Ammonium (NH_4^+)	10	< 0.04	-	1.64	-	-
(mg/L)	Iron (Fe^{2+})	11	< 0.03	-	0.20	-	-
A	Bicarbonate (HCO_3^-)	11	<0.03	-	1.80	-	-
n	Chloride (Cl^-)	12	0.50	3.88	9.80	3.12	81
i	Sulphate (SO_4^{2-})	12	0.30	1.53	4.40	1.31	85
o	Fluoride (F^-)	12	< 0.07	-	0.53	-	-
n	Nitrate (NO_3^-)	12	< 0.12	-	2.30	-	-
(mg/L)	Nitrite (NO_2^-)	12	< 0.01	-	0.07	-	-
	Fhosphates ($H_2PO_4^-$)	12	<0,13	-	< 0,15	-	-
	Silver (Ag)	10	<0.00005	-	0.00057	-	-
	Aluminium (Al)	10	0.01000	0.09404	**0.32500**	0.09621	102
	Arsenic (As)	12	<0.0002	-	0.00540	-	-
T	Boron (B)	9	0.00130	0.00497	0.01140	0.00354	71
r	Barium (Ba)	10	0.01130	0.21026	**1.82100**	0.53719	255
a	Beryllium (Be)	10	<0.00004	-	0.00007	-	-
c	Bismuth (Bi)	10	<0.00001	-	<0.00006	-	-
e	Cadmium (Cd)	10	0.00007	0.00044	0.00191	0.00054	125
	Cobalt (Co)	10	0.00003	0.00079	0.00600	0.00175	222
e	Chromium (Cr)	10	<0.0002	-	0.00076	-	-
l	Cesium (Cs)	10	<0.00002	-	0.00059	-	-
e	Copper (Cu)	11	0.00060	0.00383	**0.01120**	0.00314	82
m	Mercury (Hg)	10	<0.00003	-	0.00010	-	-
e	Manganese (Mn)	10	0.00020	0.00609	0.01970	0.00598	98
n	Molybdenum (Mo)	10	<0.00002	-	0.00037	-	-
t	Niobium (Nb)	10	<0.00001	-	0.00001	-	-
s	Nickel (Ni)	10	0.00020	0.00138	0.00450	0.00142	103
	Lead (Pb)	10	0.00017	0.00063	0.00190	0.00050	80
(mg/L)	Rubidium (Rb)	10	0.00010	0.00079	0.00280	0.00091	114
	Antimony (Sb)	10	0.00014	0.00068	0.00170	0.00055	81
	Selenium (Se)	10	<0.0003	-	0.00060	-	-
	Tin (Sn)	10	0.00003	0.00700	0.01960	0.00709	101
	Strontium (Sr)	10	0.00090	0.00486	0.01450	0.00439	90
	Tantalum (Ta)	10	<0.00001	-	<0.00006	-	-
	Tellurium (Te)	10	<0.00002	-	<0.00021	-	-
	Thallium (Tl)	10	<0.00001	-	0.00001	-	-
	Uranium (U)	10	0.00005	0.00016	0.00050	0.00013	82
	Vanadium (V)	10	0.00020	0.00098	0.00400	0.00122	124
	Yttrium (Y)	10	0.00001	0.00012	0.00030	0.00010	87
	Tungsten (W)	10	<0.0002	-	0.00280	-	-
	Zinc (Zn)	11	0.01790	0.08959	**0.16900**	0.05164	58
	Zirconium (Zr)	10	<0.00004	-	0.00053	-	-

Figure 5.
Statistics of the results of physical-chemical analysis of rainwater collected from the meteorological station near the SPS medical spa.

pH-Potentiometry, C-conductimetry, A_T-acid/base volumetry; Na^+ and K^+ by Atomic Absorption Spectrometry, the Ca^{2+}, Mg^{2+}, and Li^+, by Inductive Plasma Emission Spectrometry, Fe^{2+} by Plasma Spectrometry; Cl^-, F^- and SO_4^{2-} by Ion Chromatography, SiO_2, NO_3^-, NO_2^-, NH_4^+, and $H_2PO_4^-$ by Molecular Absorption Spectrometry; metals and Boron by Inductively Coupled Plasma Mass Spectrometry; other elements, were evaluated by calculation, as is the case of: HCO_3^-, H_T and M_T.

3. Results and interpretation

3.1 Water quality in the climatological station—Rainwater

The results of all the physical-chemical analyses of the water collected in the container associated with the climatological station are presented in detail in **Table 2**, and their results in statistical terms together with consequent bar diagrams are presented in **Figure 5**.

From the results, the very low total mineralization is highlighted, as would be expected, but the fact that they have great oscillation in their quality is emphasized, with total mineralization ranging from 2.2 to 24.4 mg/L, and the majority ions having SD_R generally greater than 80%. These situations point to the fact that its chemical composition changes over time with some significance in relative terms. Regarding the majority ionic species, the predominance of the anion Cl^- and the cations Na^+ and Ca^{2+} is noteworthy, indicating its proximity to the Atlantic Ocean.

The pH is the only parameter that apparently shows relative stability (SD_R = 9%), corresponding to acidic waters. In the trace elements, we emphasize the fact verified for the elements: Ba, Co, Cd, and V, with $SD_R \geq 124\%$. These elements may have an intermittent origin, but even so, it should be noted that they are potentially very toxic to human health. The case of Ba presents a very singular and worrying situation; overall, it presents SD_R = 255% (**Figure 5**), reaching a value of 1.8 mg/L, which corresponds to the second most representative element, after Cl^- with 2.9 mg/L (analysis of March 2007, **Table 2**). Ba is reported to be a chemical element of the alkali-earth metal family, soft, silvery in appearance, with a high melting point and toxic [13]. Ba is mainly used in spark plugs, vacuum tubes, pyrotechnic rockets, and in fluorescent lamps, besides other uses such as in barium sulphate for white pigment in paintings and in glass, and also as barium carbonate in rat poison. The negative presence of Al, Zn, Cu, Pb, among others, should also be pointed out, with Al, for example, reaching values of 0.325 mg/L, which is higher than the parametric limit for drinking water [14].

3.2 Surface runoff water quality

The results of all the physical-chemical analyses of surface runoff water collected from the simple and composite samplers are presented in detail in **Tables 3** and **4** respectively, and their results in statistical terms together with consequent bar diagrams are presented in **Figures 6** and 7, respectively.

From the outset it should be noted that in global terms there are no significant differences between the waters of the simple sampler and the composite sampler. The pH has in average terms almost the same value, with pH = 7.05 in the simple samples and pH = 7.09 in the composite samples. The total mineralization, also in average terms presents almost the same value, with M_T = 146.68 mg/L, in the plain samples

Date*		**2006 Dec**	**2007 Jan**	**2007 Feb**	**2007 Mar**	**2007 Apr**	**2007 May**	**2007 Jun**	**2007 Jul**	**2007 Oct**	**2007 Dec**	**2008 Mar**	**2008 Oct**
pH		5.40	5.55	5.81	5.28	5.93	5.41	5.62	5.21	5.23	4.87	4.74	4.17
C (μS/cm)		28.9	29.5	8.0	14.0	75.0	12.0	64.0	63.0	13.0	11.0	26.0	40.0
A_T (in HCl 0.1 N-mL/L)		<1	1.0	0.3	<0.1	—	<0.05	0.1	0.1	0.1	0.1	<0.2	<0.05
H_T (in p.p.10^5 $CaCO^3$)		1.8	4.3	0.2	0.2	2.2	0.4	1.5	0.14	—	0.2	0.3	—
Silica-SiO_2 (mg/L)		<0.3	0.4	0.2	0.3	1.1	<0.9	1.0	<0.9	<0.9	<0.9	0.4	0.9
M_T (mg /L)		13.0	14.0	7.0	6.9	—	6.7	24.4	2.2	—	3.5	11.5	—
Cations (mg/L)	Na^+	3.5	1.5	0.9	1.6	4.9	0.4	2.5	0.2	0.2	0.4	2.9	0.7
	Ca^{2+}	0.14	1.3	0.9	0.5	7.4	1.6	4.8	0.5	1.0	0.8	0.7	1.8
	K^+	0.1	0.2	0.2	0.1	1.2	0.2	0.4	<0.04	0.1	0.1	0.1	0.3
	Mg^{2+}	0.4	0.3	0.2	0.2	0.9	0.1	0.9	0.04	0.1	0.1	0.4	0.3
	NH_4^+	0.13	1.2	0.11	—	1.64	<0.04	—	—	<0.05	0.05	0.05	<0.05
	Fe^{2+}	<0.03	<0.03	<0.03	—	0.04	<0.10	0.20	<0.05	<0.05	<0.03	0.03	0.10
Anions (mg/L)	HCO_3^-	<1	1.2	1.8	<0.3	—	0.3	0.6	0.6	0.3	0.6	0.5	<0.03
	Cl^-	7.0	2.8	2.0	2.9	9.8	2.0	9.2	0.5	0.6	1.1	5.8	2.8
	SO_4^{2-}	1.2	3.1	0.5	0.8	4.4	1.2	3.6	0.3	0.7	0.5	1.1	1.0
	F^-	<0.1	<0.1	<0.07	<0.07	0.53	<0.07	0.2	<0.07	<0.07	<0.07	0.1	0.1
	NO_3^-	<0.3	2.3	0.2	0.5	1.8	0.9	1.1	<0.12	<0.12	<0.12	<0.12	<0.11
	NO_2^-	0.026	0.07	0.011	0.0	0.1	<0.01	<0.01	<0.01	<0.01	<0.01	<0.01	<0.1
	$H_2PO_4^-$	—	—	<0.13	<0.13	<0.15	<0.15	<0.15	<0.15	<0.15	<0.15	<0.15	<0.15
Trace Elements (mg/L) x10^{-3}	Ag	—	—	<0.07	0.20	<0.3	<0.2	<0.05	<0.2	<0.2	<0.1	<0.5	0.57
	Al	—	—	11	10	101	39	211	36.8	115.5	51	40.1	325
	As	<1	<1	<0.2	1.1	0.5	5.4	0.6	<0.8	0.30	<0.2	0.40	<0.69

Date*	2006 Dec	2007 Jan	2007 Feb	2007 Mar	2007 Apr	2007 May	2007 Jun	2007 Jul	2007 Oct	2007 Dec	2008 Mar	2008 Oct
B	—	—	1.80	—	8.0	3.1	11.4	1.3	3.0	2.0	4.4	9.7
Ba	—	—	33.7	1821	23.8	11.3	54.3	15.5	21.9	27.8	22.2	71.1
Be	—	—	0.01	0.02	0.04	0.01	0.05	<0.04	0.03	<0.06	<0.1	0.07
Bi	—	—	<0.01	<0.01	<0.05	<0.02	<0.03	<0.01	0.01	0.01	<0.01	<0.06
Cd	—	—	0.12	0.14	0.19	0.20	1.91	0.60	0.80	0.07	0.07	0.25
Co	—	—	0.03	0.06	6.0	0.14	0.78	0.07	0.32	0.10	0.12	0.27
Cr	—	—	<0.2	0.10	<1	0.10	0.70	<0.3	0.30	0.20	0.20	0.76
Cs	—	—	0.04	0.10	0.09	0.05	0.11	<0.02	0.06	0.10	0.03	0.59
Cu	—	4	0.60	1.50	7.60	2.10	11.20	1.10	2.20	5.20	1.20	5.40
Hg	—	—	<0.05	0.08	0.10	0.05	0.07	<0.03	<0.05	<0.05	<0.1	<0.18
Li	—	—	0.2	0.2	2.3	0.6	3.3	0.5	0.1	0.9	1.0	1.6
Mn	—	—	1.00	1.20	19.70	5.20	0.20	1.40	9.80	5.40	4.10	12.90
Mo	—	—	<0.02	<0.1	0.37	0.13	0.13	<0.3	0.03	0.02	0.01	0.07
Nb	—	—	<0.02	<0.04	0.01	<0.01	0.01	<0.02	<0.02	<0.01	<0.02	<0.06
Ni	—	—	0.20	0.30	3.30	0.80	4.50	0.50	1.00	0.40	<0.6	1.40
Pb			0.17	0.27	0.41	0.27	1.10	0.39	0.88	0.48	0.41	1.90
Rb	—	—	0.19	0.10	2.80	0.70	2.28	0.10	0.40	0.30	0.30	0.75
Sb	—	—	0.16	0.14	1.61	0.40	1.70	0.24	0.58	0.34	0.55	1.10
Se	—	—	<0.3	0.60	0.60	<0.5	<0.3	<0.8	<0.4	<0.3	<0.4	<0.4
Sn	—	—	0.03	1.02	14.60	2.69	19.60	3.10	5.28	2.42	2.91	18.30
Sr	—	—	1.40	2.00	14.50	2.90	11.80	0.90	4.90	0.90	3.20	5.10
Ta	—	—	<0.02	<0.01	<0.01	<0.01	<0.05	<0.03	<0.01	<0.01	<0.02	<0.06

DOI: http://dx.doi.org/10.5772/intechopen.107164

Date*	2006 Dec	2007 Jan	2007 Feb	2007 Mar	2007 Apr	2007 May	2007 Jun	2007 Jul	2007 Oct	2007 Dec	2008 Mar	2008 Oct
Te	—	—	<0.03	<0.04	<0.04	<0.03	<0.08	<0.02	<0.03	<0.03	<0.1	<0.21
Tl	—	—	0.01	<0.1	<0.02	<0.02	<0.05	<0.01	<0.02	0.01	<0.02	0.01
U	—	—	0.10	0.05	0.20	0.08	0.50	0.05	0.22	0.10	0.07	0.21
V	—	—	0.30	0.22	4.00	0.60	2.59	0.26	0.50	0.20	0.23	0.93
Y	—	—	0.02	0.01	0.18	0.30	0.27	0.03	0.13	0.04	0.03	0.18
W	—	—	<0.2	0.10	2.20	0.50	2.80	<0.4	<0.3	<0.2	0.30	0.21
Zn		140.0	36.9	17.9	126.3	85.0	160.1	32.4	110.0	53.3	54.6	169.0
Zr	—	—	<0.3	<0.04	0.37	<0.06	0.40	<0.10	0.30	<0.12	<0.20	0.53

*besides the dates on which rainwater was collected, other intermediate dates were also planned, but because in that period there was no rain, or in some cases technical and logistical problems occurred, the sampling was not effectively monthly sequential.

Table 2.
Chemical composition of rainwater from the meteorological station near to the SPS medical spa, over time.

Date*		2006 Oct	2007 Jan	2007 Feb	2007 Mar	2007 Jun	2007 Jul	2007 Aug	2007 Sep	2007 Oct	2007 Dec	2008 Mar
pH		7.02	7.08	6.71	6.87	6.81	6.75	7.44	7.52	7.01	7.24	7.1
C (μS/cm)		94.9	215.0	137.0	61.0	101.0	120.0	226.0	296.0	251.0	193.0	184.0
A_T (in HCl 0.1 N-mL/L)		—	—	3.2	5.4	5.9	8.2	20.8	28.7	20.1	9.0	7.7
H_T (in p.p.10^5 $CaCO^3$)		40.0	88.0	5.4	2.8	3.2	4.6	10.9	14.0	11.1	8.7	8.7
Silica-SiO_2 (mg/L)		6.8	8.0	5.5	4.3	10.4	4.9	17.0	25.2	9.3	7.6	7.3
M_T (mg/L)		86.0	165.0	99.7	56.3	84.0	100.0	211.0	291.0	223.0	153.5	144.0
Cations (mg/L)	Na^+	3.0	3.7	1.4	1.9	7.1	3.0	5.8	5.4	3.0	1.7	1.0
	Ca^{2+}	13.9	32.0	20.2	9.3	10.5	16.4	40.3	49.9	41.0	32.5	32.8
	K^+	3.2	6.1	5.5	1.7	2.0	7.7	3.8	11.5	13.1	3.6	1.8
	Mg^{2+}	1.2	1.9	0.64	0.8	1.2	1.1	2.0	3.5	2.0	1.4	1.1
	NH_4^+	0.13	<0.05	0.05	—	—	—	<0.05	<0.05	<0.05	<0.05	0.001
	Fe^{2+}	0.03	<0.03	<0.03	<0.03	0.05	<0.05	<0.05	<0.03	0.10	< 0.03	< 0.03
Anions (mg/L)	HCO_3^-	47.4	68.5	19.5	32.9	36.0	50.0	127.0	175.0	123.0	54.9	47.0
	Cl^-	3.4	7.5	1.6	1.4	8.2	2.5	6.0	6.6	5.3	1.2	4.6
	SO_4^{2-}	5.8	35.0	35.2	2.3	5.0	11.0	5.5	6.9	16.4	32.5	27.7
	F^-	0.3	0.2	0.1	0.2	< 0.12	0.2	0.35	0.3	0.2	0.2	0.1
	NO_3^-	—	2.0	9.8	1.3	3.2	2.7	2.7	4.7	7.5	17.3	20.0
	NO_2^-	—	<0.01	<0.005	<0.005	<0.001	0.03	<0.01	<0.01	<0.01	0.001	0.001
	$H_2PO_4^-$	—	—	0.2	0.18	0.26	0.44	0.23	1.03	1.30	0.66	0.52
Trace Elements (mg/L) x10^{-3}	Ag	—	—	<0.07	0.10	<0.05	0.05	0.70	0.03	< 0.02	<0.1	<0.05
	Al	—	—	10.50	55.00	8.00	48.40	35.00	26.40	8.10	7.00	4.50
	As	7.1	13.0	8.50	6.40	5.00	9.00	12.00	38.60	12.30	8.00	8.70
	B	—	—	10.6	8.00	9.80	17.5	20.9	35.0	24.0	16.0	15.8

DOI: http://dx.doi.org/10.5772/intechopen.107164

Date*	2006 Oct	2007 Jan	2007 Feb	2007 Mar	2007 Jun	2007 Jul	2007 Aug	2007 Sep	2007 Oct	2007 Dec	2008 Mar
Ba	—	—	25.0	519.0	19.3	19.5	46.5	40.0	20.0	35.2	186.0
Be	—	—	0.01	0.02	<0.02	<0.04	<0.03	0.05	<0.01	<0.06	<0.1
Bi	—	—	<0.01	0.04	<0.03	<0.01	<0.01	0.02	<0.01	<0.01	<0.01
Cd	<0.1	<0.1	0.08	11.90	0.13	0.12	0.13	0.66	0.03	0.02	0.13
Co	—	—	0.07	0.02	0.03	0.03	0.08	0.19	<0.03	0.03	0.12
Cr	—	—	0.50	2.10	0.90	3.20	1.00	0.80	0.60	0.70	1.40
Cs	—	—	0.49	0.20	0.21	0.63	0.38	0.64	0.28	0.10	0.16
Cu	5.0	2.0	4.00	2.60	1.70	3.80	1.60	9.30	3.50	2.50	1.90
Hg	—	—	<0.05	0.06	<0.05	<0.03	<0.03	0.03	<0.05	<0.05	<0.1
Li	—	—	13.8	1.5	2.5	4.6	3.3	7.4	4.5	7.7	19.4
Mn	—	—	0.40	0.40	0.30	0.40	0.30	0.30	0.57	0.20	0.30
Mo	—	—	0.77	1.20	1.09	3.10	4.00	4.25	3.70	1.35	1.72
Nb	—	—	<0.02	<0.01	<0.01	<0.02	<0.04	0.02	<0.02	<0.01	<0.02
Ni	—	—	1.00	0.40	<0.4	<0.1	<0.1	100	<1.0	<0.2	1.0
Pb			0.09	0.21	0.10	1.11	0.11	0.07	0.25	0.30	0.07
Rb	—	—	17.98	4.00	5.88	20.00	10.80	26.10	23.50	8.20	6.20
Sb	—	—	1.02	0.61	0.73	0.78	0.69	0.56	1.48	1.37	1.52
Se	—	—	<0.3	<0.6	<0.3	<0.8	<0.6	0.80	<0.4	<0.3	<0.4
Sn	—	—	0.03	0.13	0.63	0.05	0.08	0.03	0.06	0.03	0.04
Sr	—	—	50.60	20.20	32.20	38.00	79.10	144.0	76.60	82.60	69.10
Ta	—	—	<0.02	<0.01	<0.05	<0.03	<0.05	0.01	<0.01	<0.01	<0.02
Te	—	—	<0.03	<0.04	<0.08	<0.02	<0.05	0.03	<0.03	<0.03	<0.1
Tl	—	—	0.02	0.01	—	0.02	0.02	0.01	0.03	0.02	0.02

Date*	2006 Oct	2007 Jan	2007 Feb	2007 Mar	2007 Jun	2007 Jul	2007 Aug	2007 Sep	2007 Oct	2007 Dec	2008 Mar
U	—	—	0.80	1.97	0.45	1.48	11.30	8.22	2.14	1.31	0.76
V	—	—	1.60	1.14	1.46	2.06	1.90	1.30	2.20	2.80	2.67
Y	—	—	0.03	0.08	0.05	0.10	0.02	0.06	0.05	0.01	0.01
W	—	—	0.20	0.40	0.40	0.70	1.30	1.70	0.61	0.60	0.70
Zn	—	—	127.7	9.50	7.50	6.20	37.7	29.6	4.40	11.8	35.6
Zr	—	—	<0.3	<0.04	<0.1	0.12	<0.2	0.17	<0.2	<0.12	<0.2

**besides the dates on which surface runoff water was collected, other intermediate dates were also planned, but because it had not rained in that period, or in some cases technical and logistical problems occurred, the sampling was not effectively monthly sequential.*

Table 3.
Chemical composition of surface runoff water, in simple sampler, in the rainwater sewerage system of Thermas garden street, next to the SPS medical spa, over time.

DOI: http://dx.doi.org/10.5772/intechopen.107164

Date*		2006 Oct	2007 Jan	2007 Feb	2007 Mar	2007 Jun	2007 Jul	2007 Aug	2007 Sep	2007 Oct	2007 Dec	2008 Mar
pH		7.24	7.09	6.92	6.96	6.87	7.07	7.46	7.16	7.23	6.88	7.12
C (μS/cm)		115.0	221.0	136.0	76.0	106.0	134.0	190.0	239.0	290.0	205.0	182.0
A_T (in HCl 0.1 N-mL/L)		46.1	58.3	4.8	6.4	6.1	11.2	15.5	18.4	19.8	12.0	9.9
H_T (in p.p.10^5 $CaCO_3$)		48.0	92.0	5.5	3.5	3.3	5.9	8.2	10.1	12.9	9.3	8.7
SiO_2 (mg/L)		7.9	8.1	5.7	4.2	10.6	9.5	14.5	15.4	14.1	9.5	8.1
M_T (mg/L)		101.0	169.0	102.7	66.3	86.9	119.5	169.3	212.0	255.0	170.1	148.8
Cations (mg/L)	Na^+	3.2	3.5	1.5	1.9	7.6	3.2	5.4	4.7	4.5	2.2	1.2
	Ca^{2+}	16.8	34.0	20.8	12.1	11.1	21.2	30.0	36.0	45.8	34.0	32.6
	K^+	3.7	5.7	5.5	2.1	1.9	4.3	3.8	11.6	13.0	4.4	2.3
	Mg^{2+}	1.4	1.7	0.75	0.9	1.2	1.4	1.7	2.7	3.5	1.9	1.3
	NH_4^+	<0.05	0.05	<0.05	—	—	—	<0.05	<0.05	<0.05	<0.05	0.04
	Fe^{2+}	<0.03	<0.03	<0.03	<0.03	<0.05	<0.05	<0.05	<0.03	0.10	<0.03	<0.03
Anions (mg/L)	HCO_3^-	56.2	71.1	29.3	39.0	57.2	68.3	94.6	112.0	121.0	73.2	60.4
	Cl^-	3.1	7.5	1.6	1.5	8.7	2.7	5.4	5.9	6.7	1.8	3.4
	SO_4^{2-}	8.1	35.0	29.3	3.0	4.5	7.0	7.5	14.3	25.0	25.9	21.1
	F^-	0.4	0.2	0.1	0.2	<0.12	0.3	0.3	0.2	0.4	0.2	0.1
	NO_3^-	0.3	1.9	8.1	1.4	3.8	1.4	5.4	8.8	18.9	16.4	17.8
	NO_2^-	<0.01	0.023	0.01	<0.005	<0.01	<0.01	<0.01	—	<0.01	0.01	0.01
	$H_2PO_4^-$	—	—	<0.13	<0.13	0.27	0.22	0.7	1.15	1.64	0.63	0.43
Trace Elements (mg/L) x10^{-3}	Ag	—	—	<0.07	0.10	<0.05	0.03	0.05	0.05	<0.02	<0.1	<0.05
	Al	—	—	6.20	14.00	6.20	21.40	30.00	10.10	13.10	6.00	7.10
	As	40.0	11.0	8.30	7.40	6.10	18.90	13.40	30.00	18.20	10.50	8.70
	B	—	—	11.80	9.00	9.60	18.50	19.60	28.20	39.00	23.00	16.20

Date*	2006 Oct	2007 Jan	2007 Feb	2007 Mar	2007 Jun	2007 Jul	2007 Aug	2007 Sep	2007 Oct	2007 Dec	2008 Mar
Ba	—	—	15.50	300.0	12.90	12.20	23.80	29.90	21.40	20.50	22.90
Be	—	—	<0.01	0.01	<0.02	<0.04	<0.03	0.05	0.01	<0.06	<0.1
Bi	—	—	<0.01	<0.01	<0.03	<0.01	<0.01	0.02	<0.01	<0.01	<0.01
Cd	—	—	0.04	0.02	0.11	0.06	0.12	0.18	0.03	0.02	0.04
Co	—	—	0.06	0.02	0.03	0.03	0.05	0.05	0.11	0.05	0.13
Cr	—	—	0.70	1.50	0.50	2.20	1.30	0.05	2.30	1.30	1.70
Cs	—	—	0.45	0.20	0.23	0.43	40.00	0.65	0.43	0.20	0.16
Cu	5.00	2.00	1.30	1.60	1.40	2.20	3.00	4.30	4.80	2.50	1.80
Hg	—	—	<0.05	0.05	< 0.05	<0.03	<0.03	<0.02	<0.05	<0.05	<0.1
Li	—	—	13.2	1.7	2.8	2.6	2.7	5.6	4.9	7.1	15.2
Mn	—	—	0.70	0.20	0.20	0.40	0.30	0.40	0.14	0.20	0.20
Mo	—	—	0.87	1.40	0.97	5.20	3.27	4.39	5.58	1.69	1.48
Nb	—	—	<0.02	<0.01	<0.01	<0.02	<0.04	0.02	<0.02	<0.01	<0.02
Ni	—	—	0.80	<0.1	<0.4	<0.1	<0.1	<0.6	<1.0	<0.2	0.70
Pb	—	—	0.08	0.08	0.10	0.27	0.09	0.06	0.19	0.08	0.06
Rb	—	—	17.38	5.10	5.91	11.50	11.70	268.0	28.60	11.10	7.10
Sb	—	—	0.54	0.06	0.55	0.92	1.28	0.81	1.37	1.20	1.27
Se	—	—	<0.3	<0.6	<0.3	<0.8	<0.6	0.80	<0.4	<0.3	<0.4
Sn	—	—	0.03	0.06	0.39	0.03	0.36	0.04	0.05	0.03	0.07
Sr	—	—	51.90	25.50	33.10	45.60	66.30	97.70	65.30	88.20	73.40
Ta	—	—	<0.02	<0.01	<0.05	<0.03	<0.05	<0.01	<0.01	<0.01	<0.02
Te	—	—	<0.03	<0.04	<0.08	<0.02	<0.05	<0.03	<0.03	<0.03	<0.1
Tl	—	—	0.02	0.01	0.05	0.01	0.03	0.02	0.03	0.02	0.02

Date*	2006 Oct	2007 Jan	2007 Feb	2007 Mar	2007 Jun	2007 Jul	2007 Aug	2007 Sep	2007 Oct	2007 Dec	2008 Mar
U	—	—	1.00	3.37	0 52	5.49	6.40	4.71	5.87	1.82	1.09
V	< 20.0	< 20.0	1.60	1.20	1.50	2.05	3.90	2.20	2.60	2.40	2.28
Y	—	—	0.02	0.04	0.05	0.06	0.02	0.02	0.08	0.02	0.01
W	—	—	0.20	0.30	0.40	0.80	0.09	1.50	0.82	0.60	0.50
Zn	—	—	84.30	7.50	7.90	4.60	31.70	11.60	4.50	8.70	13.10
Zr	—	—	<0.3	<0.04	<0.1	<0.1	<0.2	0.17	0.20	< 0.12	<0.2

besides the dates on which surface runoff water was collected, other intermediate dates were also planned, but because it had not rained in that period, or in some cases technical and logistical problems occurred, the sampling was not effectively monthly sequential.

Table 4.
Chemical composition of surface runoff water, in the composite sampler, in the rainwater sewerage system of Thermas garden street, next to the SPS medical spa, over time.

	Parameter	N	Min	Ave	Max	SD	SD_R (%)
	pH	11	6.71	7.05	7.52	0.25	4
	Conductivity - C (μS/cm)	11	61.00	170.81	296.00	70.19	41
	Total Alkalinity - A_T (in HCl 0.1N - mL/L)	9	3.20	12.11	28.70	8.31	69
	Total Hardness - H_T (in p.p.10^5 $CaCO_3$)	11	2.80	17.95	88.00	24.23	135
	Sílica - SiO_2 (mg/L)	11	4.30	9.66	25.20	5.92	61
	Total Mineralization-M_T (mg /L)	11	56.30	146.68	291.00	68.32	47
C	Sodium (Na^+)	11	1.00	3.36	7.10	1.88	56
a	Calcium (Ca^{2+})	11	9.30	27.16	49.90	13.15	48
t	Potassium (K^+)	11	1.70	5.45	13.10	3.71	68
i	Magnesium (Mg^{2+})	11	0.64	1.53	3.50	0.77	50
o	Lithium (Li^+)	9	0.0015	0.0072	0.0194	0.0055	77
n	Ammonium (NH_4^+)	8	<0.05	-	0.13	-	-
(mg/L)	Iron (Fe^{2+})	11	<0.03	-	0.10	-	-
A	Bicarbonate (HCO_3^-)	11	19.50	71.02	175.00	46.54	66
n	Chloride (Cl^-)	11	1.20	4.39	8.20	2.41	55
i	Sulphate (SO_4^{2-})	11	2.30	16.66	35.20	12.67	76
o	Fluoride (F^-)	11	<0.12	-	0.35	-	-
n	Nitrate (NO_3^-)	10	1.30	7.12	20.00	6.30	89
(mg/L)	Nitrite (NO_2^-)	10	<0.001	-	0.03	-	-
	Fhosphates ($H_2PO_4^-$)	9	0.18	0.54	1.30	0.37	70
	Silver (Ag)	10	<0.00005	-	0.00070	-	-
	Aluminium (Al)	9	0.00450	0.02254	**0.05500**	0.01833	81
	Arsenic (As)	11	0.00500	0.01169	0.03860	0.00884	76
T	Boron (B)	9	0.00800	0.01751	0.03500	0.00790	45
r	Barium (Ba)	9	0.01930	0.10117	**0.51900**	0.15586	154
a	Beryllium (Be)	9	<0.00001	-	0.00005	-	-
c	Bismuth (Bi)	9	<0.00001	-	0.00004	-	-
e	Cadmium (Cd)	11	<0.0001	-	**0.01190**	-	-
	Cobalt (Co)	9	<0.00003	-	0.00019	-	-
e	Chromium (Cr)	9	0.00050	0.00124	0.00320	0.00083	67
l	Cesium (Cs)	9	0.00010	0.00034	0.00064	0.00019	56
e	Copper (Cu)	11	0.00160	0.00345	0.00930	0.00212	62
m	Mercury (Hg)	9	<0.00003	-	0.00006	-	-
e	Manganese (Mn)	9	0.00020	0.00035	0.00057	0.00010	28
n	Molybdenum (Mo)	9	0.00077	0.00235	0.00425	0.00131	56
t	Niobium (Nb)	9	<0.00001	-	0.00002	-	-
s	Nickel (Ni)	9	<0.0001	-	**0.10000**	-	-
(mg/L)	Lead (Pb)	9	0.00007	0.00026	0.00111	0.00031	122
	Rubidium (Rb)	9	0.00400	0.01363	0.02610	0.00787	58
	Antimony (Sb)	9	0.00056	0.00097	0.00152	0.00036	37
	Selenium (Se)	9	<0.0003	-	0.00080	-	-
	Tin (Sn)	9	0.00003	0.00012	0.00063	0.00018	152
	Strontium (Sr)	9	0.02020	0.06582	**0.14400**	0.03482	53
	Tantalum (Ta)	9	<0.00001	-	0.00001	-	-
	Tellurium (Te)	9	<0.00002	-	0.00003	-	-
	Thallium (Tl)	8	0.00001	0.00002	0.00003	0.00001	32
	Uranium (U)	9	0.00045	0.00316	**0.01130**	0.00364	115
	Vanadium (V)	9	0.00114	0.00190	0.00280	0.00055	29
	Yttrium (Y)	9	0.00001	0.00005	0.00010	0.00003	65
	Tungsten (W)	9	0.00020	0.00073	0.00170	0.00045	61
	Zinc (Zn)	9	0.00440	0.03000	**0.12770**	0.03669	122
	Zirconium (Zr)	9	<0.00004	-	0.00017	-	-

Figure 6.
Statistics of the results of physical-chemical analysis of water collected in the simple sampler, from surface runoff water at Thermas garden street, near the SPS medical spa, at different times of the year.

	Parameter	N	Min	Ave	Max	SD	SD_R (%)
pH		11	6.87	7.09	7.46	0.17	2
Conductivity - C (μS/cm)		11	76.00	172.18	290.00	61.71	36
Total Alkalinity - A_T (in HCl 0.1N - mL/L)		11	4.80	18.95	58.30	16.55	87
Total Hardness - H_T (in p.p.10^5 of $CaCO_3$)		11	3.30	18.85	92.00	26.01	138
Sílica - SiO_2 (mg/L)		11	4.20	9.78	15.40	3.44	35
Total Mineralization – M_T (mg /L)		11	66.30	145.51	255.00	54.22	37
C	Sodium (Na^+)	11	1.20	3.54	7.60	1.82	52
a	Calcium (Ca^{2+})	11	11.10	26.76	45.80	10.54	39
t	Potassium (K^+)	11	1.90	5.30	13.00	3.52	66
i	Magnesium (Mg^{2+})	11	0.75	1.68	3.50	0.76	45
o	Lithium (Li^+)	9	0.0017	0.0062	0.0152	0.0046	74
n	Ammonium (NH_4^+)	8	<0.05	-	0.05	-	-
(mg/L)	Iron (Fe^{2+})	11	<0.03	-	0.10	-	-
A	Bicarbonate (HCO_3^-)	11	29.30	69.30	121.00	28.53	41
n	Chloride (Cl^-)	11	1.50	4.39	8.70	2.44	55
i	Sulphate (SO_4^{2-})	11	3.00	16.43	35.00	10.71	65
o	Fluoride (F^-)	11	< 0.12	-	0.40	-	-
n	Nitrate (NO_3^-)	11	0.30	7.65	18.90	6.69	87
(mg/L)	Nitrite (NO_2^-)	10	<0.005	-	0.023	-	-
	Phosphates ($H_2PO_4^-$)	9	<0.13	-	1.64	-	-
	Silver (Ag)	9	<0.00002	-	0.00010	-	-
	Aluminium (Al)	9	0.00600	0.01268	0.03000	0.00778	61
	Arsenic (As)	11	0.00610	0.01568	0.04000	0.01014	65
T	Boron (B)	9	0.00900	0.01943	0.03900	0.00911	47
r	Barium (Ba)	9	0.01220	0.05101	**0.30000**	0.08819	173
a	Beryllium (Be)	9	<0.00001	-	0.00005	-	-
c	Bismuth (Bi)	9	<0.00001	-	0.00002	-	-
e	Cadmium (Cd)	9	0.00002	0.00007	0.00018	0.00005	76
	Cobalt (Co)	9	0.00002	0.00006	0.00013	0.00004	60
e	Chromium (Cr)	9	0.00005	0.00128	0.00230	0.00071	56
l	Cesium (Cs)	9	0.00016	0.00475	0.04000	0.01246	262
m	Copper (Cu)	11	0.00130	0.00272	0.00500	0.00131	48
e	Mercury (Hg)	9	<0.00002	-	0.00005	-	-
n	Manganese (Mn)	9	0.00014	0.00030	0.00070	0.00017	54
t	Molybdenum (Mo)	9	0.00087	0.00276	0.00558	0.00177	64
s	Niobium (Nb)	9	<0.00001	-	0.00002	-	-
	Nickel (Ni)	9	<0.0001	-	0.00080	-	-
(mg/L)	Lead (Pb)	9	0.00006	0.00011	0.00027	0.00007	60
	Rubidium (Rb)	9	0.00510	0.04071	**0.26800**	0.08064	198
	Antimony (Sb)	9	0.00006	0.00089	0.00137	0.00042	47
	Selenium (Se)	9	<0.0003	-	0.00080	-	-
	Tin (Sn)	9	0.00003	0.00012	0.00039	0.00014	117
	Strontium (Sr)	9	0.02550	0.06078	**0.09770**	0.02272	37
	Tantalum (Ta)	9	<0.00001	-	<0.00005	-	-
	Tellurium (Te)	9	<0.00002	-	<0.0001	-	-
	Thallium (Tl)	9	0.00001	0.00002	0.00005	0.00001	49
	Uranium (U)	9	0.00052	0.00336	0.00640	0.00219	65
	Vanadium (V)	9	0.00120	0.00219	0.00390	0.00074	34
	Yttrium (Y)	9	0.00001	0.00004	0.00008	0.00002	62
	Tungsten (W)	9	0.00009	0.00058	0.00150	0.00040	69
	Zinc (Zn)	9	0.00450	0.01932	**0.08430**	0.02425	126
	Zirconium (Zr)	9	<0.00004	-	0.00017	-	-

Figure 7.
Statistics of the results of physical-chemical analysis of water collected in the composite sampler, from surface runoff water at Thermas garden street, near the SPS medical spa, at different times of the year.

and M_T = 145.51 mg/L in the composite samples. The majority ionic composition in average terms, in both situations the three main cations and anions follow the same order, that is: $Ca^{2+} >> K^+ > Na^+$, and $HCO_3^- > SO_4^{2-} > NO_3^-$, respectively. In relation to trace elements, in both situations Ba and Sr. appear as predominant elements, and others with relatively significant values such as Rb, Zn, B, Al and As.

When comparing the two sets in terms of simple and composite samples, for the maximum values, it is worth noting that there are some slight differences, as for example in the M_T. The waters from the simple samples are slightly more mineralized (M_T = 291.0 mg/L) than the waters from the composite samples (M_T = 255 mg/L). Note that the simple samples are a consequence of the water from the first rainfall while the composite samples correspond to water not only from the first rainfall but from others collected over time. In principle, the first runoff tends to be more mineralized, as they are the ones that first come into contact with heavily polluted surfaces.

From the point of view of the stability of the various chemical elements in this type of water over time, significant changes can be noted from sample to sample; the majority of the elements, both major and trace, present SD_R greater than 50%, showing great changes over time. The most serious situations were observed for Ba, Sn and Zn in both types of samples (SD_R of 117 to 173%), as well as for Pb and U in the simple samples with SD_R of 122 and 115%, respectively, and finally, the cases of Cs and Rb, in the composite samples, with SD_R of 262 and 198%, respectively. In addition to the previous situation, there are cases in which it was not possible to perform total statistics because in most of the surveys some elements appear below the detection limit, and occasionally occur with relatively significant values, examples being Ni and Cd in the simple samples.

In comparative terms with rainwater quality, the following should be noted at the outset: (i) surface runoff water has a much higher pH than rainwater, as the latter have an essentially acidic character ($pH_{ave\text{-}rainwater}$ = 5.27), and surface runoff water have much higher values ($pH_{ave\text{-}surface\text{-}runoff} \cong 1.34 pH_{ave\text{-}rainwater}$ = 7.06), with a neutral character; (ii) surface runoff water has a much higher total mineralization than rainwater, which in average terms is: $M_{T\text{-}ave\text{-}surface\text{-}runoff} \cong 14.75 M_{T\text{-}ave\text{-}rainwater}$ = 146.2 mg/L; (iii) rainwater has almost zero values for Total Alkalinity (A_T), for silica (SiO_2), and even for Total Hardness (H_T), while surface runoff water reaches levels with some significance (maximum values: A_T = 58.3 0 mL/L, SiO_2 = 25.2 mg/L, H_T = 92.0 p.p.10^5 of $CaCO_3$; (iv) in rainwater, the elements of the major ionic component are: Cl^- in the anion, and Na^+ and Ca^{2+} in the cations; in surface runoff water the anions and cations, respectively, predominate in the following elements: $HCO_3^- > SO_4^{2-} > NO_3^-$; and $Ca^{2+} >> K^+ > Na^+$, emphasizing that Cl^- in particular ceases to have meaning and Na^+, never appears as the predominant cation; (v) Phosphate ion ($H_2PO_4^-$) in rainwater never appeared and in surface runoff water became part of the ensemble of anions observed; (vi) regarding the trace elements, it is emphasized the fact that the most representative element in rainwater was Ba, reaching a maximum value of 1.8 mg/L. In surface runoff water, it also continues to be the most representative element, although with slightly lower values, with maximum values of 0.52 mg/L and 0.30 mg/L, in simple and composite samples, respectively; (vii) regarding still other trace elements, as in rainwater, several chemical elements are still present, with particular attention to Zn, which in rainwater reaches 0.169 mg/L and in the simple sample of the surface runoff water reaches 0.130 mg/L; and (viii) about trace elements in surface runoff water, it is also worth noting the fact that these include with some significance Sr. (0.144 mg/L) and Ni (0.10 mg/L) in simple samples, and Rb (0.268 mg/L) in the composite samples, being noteworthy that these elements are not as important in rainwater.

3.3 Comparison between the quality of natural mineral water from the traditional spring and the quality of rainwater and surface runoff water

The chemical quality of NMW from the Traditional Spring (NT) of the SPS medical spa, was already studied by Lepierre in 1903 and 1928, at least in the global parameters and majority ions [5]. However, it was from 1985 onwards that a systematic control over time was carried out, as mentioned in the methodology section. The results obtained over 29 years (1985 to 2014) are presented in statistical terms in **Table 1**. These are in the basis of the present work because as already mentioned in item 1, this systematic control allowed us to verify, in 2005–2006, unusual results of Pb, Zn and Cu, leading to subsequent work, namely the study of the quality of rainwater and surface runoff water, to identify, and annul, or minimize the problem.

From a global point of view, it should be noted that the NMW of the NT (**Table 1**), in a universe of 96 samples over time: the M_T and pH have average values of 359.68 mg/L, and 8.82, respectively, with only SD_R of 2 and 1%, respectively. The rainwater, for those parameters, in average terms, was obtained: M_T = 9.91 mg/L and pH = 5.27, with SD_R of 64 and 9% respectively. The surface runoff water, also in average terms, for the simple samples: M_T = 146.68 mg/L and pH = 7.05, with SD_R of 47 and 4% respectively, and for the composite samples: M_T = 145.51 mg/L and pH = 7.09, with SD_R of 37 and 2% respectively.

It is emphasized the fact that, in global terms, the silica component (in the non-ionized form), in average terms, in NT, presents a SiO_2/M_T ratio = 19%, and with excellent stability (SD_R - 5%), while in rainwater that ratio is null, and in surface runoff water it is about 7%, with very large SD_R. Regarding the majority ions, the NT waters always present HCO_3^- and Na^+ as the most representative anion and cation, respectively, and very stable over time; in this water, the predominance of the ions, are according to the following order, from most to least representative: $HCO_3^- >> Cl^- > F^- > SO_4^{2-}$, for the anions, and $Na^+ >> K^+ > Ca^{2+}$ for the cations. In rainwater, the predominant anions, in average terms, are Cl^- and SO_4^{2-}, although occasionally Cl^- has a lower content than SO_4^{2-}; it should be noted that in these waters the HCO_3^- and F^- anions almost always have negligible contents. Still in rainwater, in terms of cations, the predominant ones are Na^+ and Ca^{2+}, almost with similar levels, and with some significance in relative terms the K^+ and Mg^{2+}.

In the case of surface runoff water in simple and composite samples, HCO_3^- is always the main anion, as it happens in NT waters; on average, $HCO_3^- >> SO_4^{2-} > NO_3^- > Cl^-$, although occasionally, except for HCO_3^-, the sequence is not always the same; still concerning anions, F^- is always present in NT waters, with an average value of 17.71 mg/L, and corresponding SD_R of 3%, and in surface runoff water appears almost without expression, having found the maximum value of F^- = 0.4 mg/L. Still in the case of surface runoff water in simple and composite samples, for the cations in average terms it was verified: $Ca^{2+} > K^+ > Na^+ > Mg^{2+}$, being of note that in the simple samples in fact Ca^{2+} was always the one with the highest content, being that the order of the other elements, in the individual analyses, was not always the same; in the composite samples, that order is always maintained in all the individual samples. Regarding the NT water, the first three cations are also predominant in this water, although in the inverse order ($Na^+ >> K^+ > Ca^{2+}$), being worth mentioning that the NT water does not present Mg^{2+}. Regarding trace elements, the NMW of the NT has a typical ensemble, which is acquired in the extensive path of the water particles, which at the beginning are rainwater, infiltrate the rock mass, and evolve over many kilometers, passing even, along some extensive veins, which are sometimes rich in singular minerals, namely metals. According to the research already carried out (**Table 1**), the elements that are understood to form part of this ensemble, from the most abundant to the least abundant (using the average content as a reference), are presented in **Table 5**.

It is admitted that other elements, never surveyed so far, may occur, such as, for example, gold (Au). The case of Br was surveyed only three times and always appeared; to confirm order 3, many more surveys will have to be carried out, and to verify that the SD_R is less than 10%. The case of Al appears with a SD_R = 217%, which means that, besides being an element of the typical ensemble of NMW, contamination sometimes occurs because of anthropic actions. A less serious situation is found with the As and the Mn; there is a need to do more research on these elements to ascertain their order, without considering the elements with added anthropic component. The situation of Pb, Zn and Cu, surveyed many times in the studies over time (**Table 1**), are understood not to be part of the ensemble of natural trace elements of the NMW of the NT, as they were recorded many times below the limit of detection (l.d.) as can be seen in **Table 1** and especially in **Table 6**, where the cases of Pb and Cu, 43% of the times surveyed, the result was below the l.d., and Zn only 11% below the l.d., as it is often affected by anthropic actions. Other trace elements that have sometimes been recorded in the NMW of NT, and that in the 2005–2006 season were not of concern, but with the detailed analysis in the present chapter, one gets the notion that there is a great possibility of their occurrence being a consequence of anthropic actions; they are in particular: Ba, Ni, V, Cd, Co and U, since in some cases most of the times situations below the l.d. have been verified (**Table 6**).

For a joint analysis of the trace elements obtained in rainwater and surface runoff water, the chemical elements studied were organized according to the dosages obtained in the NMW of the NT, resulting in **Table 7**, with the elements organized into two large groups:

- *Group I*, which includes the chemical elements understood to be *part of the ensemble of the natural trace elements* of NMW. Crossing the results, of the three

Element	Order	N	min	ave	max	SD	SD_R (%)
Boron (B)	1	51	0.3600	0.4263	0.4760	0.0245	6
Bromide (Br)	2	3	0.1300	0.1750	0.2600	0.0601	34
Tungsten (W)	3	53	0.0370	0.0791	0.0990	0.0101	13
Strontium (Sr)	4	50	0.0570	0.0667	0.0730	0.0032	5
Cesium (Cs)	5	42	0.0460	0.0624	0.0721	0.0043	7
Rubidium (Rb)	6	42	0.0530	0.0590	0.0670	0.0028	5
Aluminium (Al)	7	51	0.0021	0.0437	0.6400	0.0950	217
Arsenic (As)	8	53	0.0030	0.0045	0,0180	0.0022	49
Manganese (Mn)	9	52	0.0013	0.0020	0.0060	0.0009	45
Beryllium (Be)	10	53	0.0003	0.0006	0.0014	0.0002	28

from **Table 1; N: number of samples studied, min: minimum, ave.: average, SD: standard deviation, SD_R: relative standard deviation = (SD/ave)x100. Order: 1 is the most abundant, 10 is the least abundant.*

Table 5.
Main trace elements from chemical analyses carried out between 1985 and 2014 of NMW of the NT.*

Ele.	N	n	l.d. (%)	l.d. (mg/L)	max (mg/L)	Observations
Pb	51	22	43	0.00006	0.0730	max was in 2005/3; from 2006/11 rn
Zn	53	6	11	0.0002	0.2270**	max occurred in 2005/3; from 2007/7 rn
Cu	51	22	43	0.00005	0.0550	max was observed in 2005/3; from 2008/6 rn
Al	51	0	0	—	0.640***	max was observed in 1989/10; it is admitted that from 1998/6 rn
Ba	53	6	11	0.0003	0.0063	max was observed in 2014/4
Ni	52	30	58	0.0002	0.0150	max occurred in 2005/3; from 2005/7 rn
V	53	27	51	0.0002	0.0009	max was observed in 2003/11; in 2005/8 rn
Cd	34	5	15	0.00002	0.0006	max was observed in 2012/1; in 2012/4 rn
Co	53	34	64	0.00001	0.0003	max occurred in 2005/3; in 2005/7 rn
U	47	27	57	0.00002	0.0001	max occurred in 2005/12; in 2006/2 rn

**from Table 1; l.d.: detection limit, l.d (%) = (n/N) × 100, N: number of samples studied, n: number of samples with results < l.d., max: maximum, rn: returned to normal.*
***note: in 2003, Zn = 0.3140 mg/L was punctually observed.*
****it is admitted that NMW, at its origin, has Al contents estimated at around 0.0021 mg/L.*

Table 6.
Detailed aspects of the occurrence of some trace elements in the NMW of the NT.*

types of waters, it is only extracted the clear notion, that the Al, in fact, could be a consequence of relations with the rainwater and surface runoff waters. Such an inference is the result of the high SD_R in the NT water, and the relative abundance of this element sometimes in rainwater and surface runoff water. It is admitted that As and Mn also have some interference with rainwater and surface water. The case of Br, with almost no information, should be investigated in the future in all water types.

- *Group II*, includes the chemical elements that is admitted *not do part of the ensemble of the natural trace elements* of NMW, or if they are part of the natural elements, their content is very low and close to the detection limit. This understanding results from the number of times they appear below the detection limit, and on the other hand by the significant quantities that at least some appear in rainwater and surface runoff water. It is important to mention that the occurrence of organic and other soils, which have been removed with the improvement works, around the NT will also contribute to the addition of some elements. Thus, in this group, it was subdivided into three subgroups: (i) *Sub-group II-a,* in which the chemical elements sometimes occur with *some significance*; it is understood that there will be a relationship with some elements in rainwater and surface runoff water, namely Zi, Cu, and particularly Ba, in addition to other elements in lesser influences such as Pb, Ni, and Sn; (ii) *Sub-group II-b*, includes the chemical elements, where they sometimes occur with *slight significance*; it is understood that there will be a relationship with some elements in rainwater and surface runoff water, namely in V, Cd, Co, Hg, and even U; (iii) *Sub-group II-c*, includes chemical elements where they sometimes occur with *very slight significance*; it includes the elements Ta, Y, and Ag; the occurrence of these elements admitted that may have nothing to do with rainwater and surface runoff water.

Group	Element	Traditional Springer—NT				Rainwater				Surface runoff water**			
		N	ave	max	SD_R(%)	N	ave	max	SD_R (%)	N	ave	max	SD_R (%)
I	B	51	0.4263	0.4760	6	9	0.0050	0.0114	71	9	0.0194	0.03900	47
	Br	3	0.1750	0.2600	34	—	—	—	—	—	—	—	—
	W	53	0.0791	0.0990	13	10	—	0.00280	—	9	0.0007	0.00170	61
	Sr	50	0.0667	0.0730	5	10	0.0049	0.01450	90	9	0.0658	0.14400	53
	Cs	42	0.0624	0.0721	7	10	—	0.00280	—	9	0.0048	0.04000	262
	Rb	42	0.0590	0.0670	5	10	0.0008	0.00280	114	9	0.0407	0.26800	198
	Al	51	0.0437	**0.640**	**217**	10	0.0940	**0.32500**	102	9	0.0225	0.05500	81
	As	53	0.0045	0,0180	49	10	—	0.00540	—	11	0.0157	**0.0400**	65
	Mn	52	0.0020	0.0060	45	10	0.0061	0.01970	98	9	0.0003	0.00070	54
	Be	53	0.0006	0.0014	28	10	—	0.00007	—	9	—	0.00005	—
II-a	Zn	53	—	0.3140	—	11	0.0896	**0.16900**	58	9	0.0300	0.12770	122
	Pb	52	—	0.0700	—	10	0.0006	0.00190	80	9	0.0003	0.00111	122
	Cu	51	—	0.0550	—	11	0.0038	**0.01120**	82	11	0.0035	0.00930	62
	Ni	52	—	0.0150	—	10	0.0014	0.00450	103	9	—	0.10000	—
	Cr	53	—	0.0124	—	10	—	0.00076	—	9	0.0012	0.00320	67
	Sb	53	—	0.0070	—	10	0.0007	0.00170	81	9	0.0010	0.00152	37
	Ba	53	—	0.0063	—	10	0.2103	**1.82100**	255	9	0.1012	**0.5190**	154
	Mo	53	—	0.0060	—	10	—	0.00037	—	9	0.0028	0.00558	64
	I	3	—	0.0020	—	—	—	—	—	—	—	—	—
	Tl	41	—	0.0015	—	10	—	0.00001	—	9	0.00002	0.00005	49
	Se	49	—	0.0013	—	10	—	0.00060	—	9	—	0.00080	—
	Sn	53	—	0.0010	—	10	0.0070	0.01960	101	9	0.00012	0.00063	152

DOI: http://dx.doi.org/10.5772/intechopen.107164

Group	Element	Traditional Springer—NT				Rainwater				Surface runoff water**			
		N	ave	max	SD_R(%)	N	ave	max	SD_R (%)	N	ave	max	SD_R (%)
II-b	V	53	—	0.0009	—	10	0.0010	0.00400	124	9	0.00219	0.00390	34
	Cd	53	—	0.0006	—	10	0.0004	0.00191	125	9	—	0.01190	—
	Co	53	—	0.0003	—	10	0.0008	0.00600	222	9	—	0.00019	—
	Hg	47	—	0.0003	—	10	—	0.00010	—	9	—	0.00006	—
	Te	48	—	0.0003	—	10	—	<0.00021	—	9	—	0.00003	—
	Zr	41	—	0.0002	—	10	—	0.00053	—	9	—	0.00017	—
	Bi	40	—	0.0001	—	10	—	<0.00006	—	9	—	0.00004	—
	Nb	52	—	0.0001	—	10	—	0.00001	—	9	—	0.00002	—
	U	47	—	0.0001	—	10	0.0002	0.00050	—	9	0.0032	0.01130	115
II-c	Ta	41	—	0.00004	—	10	—	<0.00006	—	9	—	0.00001	—
	Y	53	—	0.00002	—	10	0.0001	0.00030	87	9	0.00005	0.00010	65
	Ag	49	—	<0.0005	—	10	—	0.00057	—	9	—	0.00070	—

*from **Table 1** and **Figures** 5–7.
**the present values were the ones that occur in greater dosage, among those obtained in the simple and composite samples; N: number of samples studied, ave.: average, max: maximum, SD_R: relative standard deviation.

Table 7.
Comparison of trace element levels recorded in NMW from the NT of the SPS medical spa, with those recorded in local rainwater and surface runoff water.*

4. Conclusions

The natural mineral water (NMW) from the Traditional Spring (NT) of the São Pedro do Sul medical spa is fundamental to the public health of a considerable group of people who use that medical spa. As a result, the local economy, through direct revenues from thermal treatments and other indirect revenues from tourism, hotel, catering, and other activities, is very important to boost the economy of the entire region where the medical spa is located. The change in the chemical quality of NT water leads to the assumption that there is contamination and, as a result, the health authorities impose the closure of the use of that water. Therefore, the chemical composition of the water is analyzed systematically over time. Such quality control allows checking if anomalies are occurring, and sometimes their causes are enigmatic.

Infiltration of rainwater, or even of surface runoff water in the proximity of the NT has the potential to interfere with the water quality of the NT. There is also the possibility that other causes may be at the origin of these potential problems, such as the migration of chemical elements from the garden soils surrounding the NT house; these latter situations are not studied in this chapter. It is therefore on the chemical quality of rainwater and surface runoff water from the medical spa area, subjected to detailed physical-chemical analyses for about two years, that this chapter focuses, in order to try to understand if, from these may result in changes in the quality of the NMW of the NT.

Regarding rainwater (**Figure 5**), total mineralization (M_T) shows a variation from 2.2 to 24.4 mg/L and pH between 4.17 and 5.93. In relation to the majority ionic species, the predominance of the anion Cl^- (0.5 to 9.8 mg/L) and the cations Na^+ (0.2 to 4.9 mg/L) and Ca^{2+}(0.14 to 7.4 mg/L) is highlighted, denouncing its proximity to the Atlantic Ocean, which is only about 55 km to the west. In the trace elements, emphasis is given to the fact that elements Ba, Co, Cd, and V were found, with a relative standard deviation, SD_R, greater than 120%, pointing to intermittent origins; the case of Ba, presents a very singular and worrying situation, with SD_R = 255%, reaching a content of 1.82 mg/L, which corresponded to the second most representative element, after Cl^- with 2.9 mg/L (analysis March 2007, **Table 2**). It is also noted the almost systematic presence of other trace elements, such as Al, Zn, Cu, and Pb, and it should be emphasized that, for example, Al reaches values of 0.325 mg/L, which is higher than the parametric limit of water for human consumption.

About the surface runoff waters, which are the result of rain that falls in the garden area and street surrounding the NT site, at a maximum distance of about 250 m from the sampling site (**Figure 3**), it should be noted that there are no major differences in the chemical quality of the samples of the 1st rainfall (simple samples—**Figure 6**) with the samples collected in various moments during the rains (composite sample—**Figure 7**). From the point of view of the stability of the various chemical elements in this type of water over time, there are significant changes from sample to sample; most elements, whether majority or vestigial, have an SD_R greater than 50%, showing large changes over time. In comparative terms with rainwater quality, the following aspects are emphasized:

i. surface runoff waters have a higher M_T and pH than rainwater; in average terms, it is verified: (a) $M_{T\text{-ave-surface-runoff}} \cong 14.75 \text{x} M_{T\cdot\text{ave-rainwater}}$ = 146.2 mg/L; and (b) $pH_{\text{ave-surface-runoff}} \cong 1.34 \text{ x } pH_{\text{ave-rainwater}}$ = 7.06;

ii. the surface runoff waters present contents with some significance of Total Alkalinity (A_T), silica (SiO_2), and even for Total Hardness (H_T), with maximum values: A_T = 58.3 0 mL/L, SiO_2 = 25.2 mg/L, and H_T = 92.0 p.p.10^5 of $CaCO_3$, while in rainwater these parameters show almost zero values;

iii. on the elements of the majority ionic component while rainwater presents Cl^- in the anions, and Na^+ and Ca^{2+} in the cations, in surface runoff water predominate in the anions: $HCO_3^- > SO_4^{2-} > NO_3^-$, and in the cations: $Ca^{2+} >> K^+ > Na^+$, emphasizing that in particular Cl^- ceases to have significance and Na^+, never appears as the predominant cation; it is also mentioned that Phosphate ($H_2PO_4^-$) in rainwater never appeared and in surface runoff water became part of the set of anions observed;

iv. regarding the trace elements, it is emphasized that the most representative element, as in rainwater, continues to be Ba, despite having lower values, with a maximum value of 0. 52 mg/L; in relation to other trace elements, such as in rainwater, several chemical elements continue to be present, with particular attention to Zn, which in rainwater reaches 0.169 mg/L and in the simple surface runoff water sample it reaches reach 0.130 mg/L; on trace elements, in surface runoff water, it is also worth noting the fact that they include with some significance Sr (0.144 mg/L), Ni (0.10 mg/L), and Rb (0.268 mg/L), being of mention that these elements are not so important in rainwater.

On the NMW of the NT, that being a groundwater resulting from an extensive hydrogeological circuit, with its beginning as rainwater, which infiltrates at depth, and as it percolates along the essentially granite rock mass, in the various water/rock interactions, it acquires its own chemical composition. From a global point of view, based on the universe of 96 physical-chemical analyses carried out over time (**Table 1**), it is noted that it has very good stability, particularly evident in the global physical-chemical parameters and the majority ionic component, which generally present SD_R of less than 10%. M_T and pH present mean values of 359.68 mg/L, and 8.82, respectively. It is emphasized that these are alkaline waters, while rainwater is acidic and surface runoff water is neutral. Still in the global parameters, it is emphasized the occurrence in the water of the NT, of: (i) Sulphuration, with average values of 21.3 mL/L, which is a parameter that corresponds to the total ionic water content of S_2^- specie, including simple and complex sulphury forms [15], and does not appear in the other waters; (ii) Non-ionized silica, with average value of 67.94 mg/L, which corresponds to a considerable representation in relation to the total mineralization (SiO_2/M_T = 19%); the silica, is almost null in rainwater, and in surface runoff water presents SiO_2/M_T = 7%. In relation to the majority ions, systematically, the most representative ions, from highest to lowest, are: $HCO_3^- >> Cl^- > F^- > SO_4^{2-}$, for the anions, and $Na^+ >> K^+ > Ca^{2+}$ for the cations; these situations are completely different from those seen in rainwater and surface runoff water. We point out the particularity of F^- that in the NT water occurs always, with an average value of 17.71 mg/L, which is relatively considerable, while in rainwater and surface runoff water, this element appears almost without expression, with the maximum value of F^- = 0.53 mg/L in rainwater. In relation to the trace elements in the NT water and comparison with what occurs in rainwater and surface run-off water, the elements are organized according to the following (**Table 7**):

- *Group I*: B > Br > W > Sr. > Cs > Rb > Al > As>Mn > Be, which correspond to the most systematically representative chemical elements and are understood to be *part of the ensemble of natural trace elements* of NMW. These elements were always quantified whenever they were surveyed. Some present excellent SD_R, less than 10%, as is the case of B, Sr., Cs and Rb. We highlight the case of Al, which presents an SD_R = 217%, therefore sometimes with relatively high values, which is understood to be a punctual consequence of mixtures and/or infiltration of rainwater and surface runoff water; it is also admitted that As and Mn also have some interference with rainwater and surface runoff water.

- *Group II*, which includes the chemical elements that are admittedly *not part of the ensemble of natural trace elements* of NMW, or if they are part of the natural elements, their content is very low and close to the detection limit; this group is subdivided into three subgroups: (i) *Sub-group II-a:* Zn > Pb > Cu > Ni > Cr > Sb > Ba>Mo > I > Tl > Se > Sn, in which chemical elements sometimes occur with *some significance*; it is understood that there will be a relationship with some elements from rainwater and surface runoff, notably in the cases of Zi, Cu and particularly Ba, in addition to other elements in lesser influences such as Pb, Ni and Sn; (ii) *Sub-group II-b*: V, Cd, Co, Hg, Te, Zr, Bi, Nb, U, in which the chemical elements sometimes occur with *slight significance*; it is understood that there will be a relationship with some elements of rainwater and surface runoff water, namely in V, Cd, Co, Hg and even U; (iii) *Sub-group II-c*: Ta, Y, Ag, in which the chemical elements occur sometimes with *very slight significance*; their occurrence admittedly may have nothing to do with rainwater and surface runoff water.

The present work was triggered by the occurrence of abnormal levels of Zn, Cu and Pb in March 2005; however, when analyzing in detail the chemical situation during the 29 years of analyses of the water of the NT, occasionally there were problems with other elements, which have since been resolved, as shown in **Table 6**. The main action taken to solve these problems was the removal of organic and other soils from around the NT house, replacing them with impermeable clays, among others. Attention should continue to be paid to the monitoring of all the elements that constitute the External Monitoring System, namely the recording of the quality of groundwater from the piezometers, and, as far as possible, measures should be taken to eliminate all the traditional garden area, replacing it by situations that favor impermeable zones. Solutions should also be found to cancel traffic and parking in the area near the NMW abstractions.

Finally, we emphasize the occurrences of anomalous chemical elements from the domain of trace elements in rainwater, namely those that occur systematically, such as Ba, Al, Zn, Cu, Pb, Cd, Co, Mn, Ni, Rb, Sb, Sn, Sr., U, V and I. It will be important in the future to continue researching these chemical elements in rainwater, and to simultaneously study the direction of local and regional winds, in order to find out the origin of pollution from the air, as these chemical elements could be harmful to the NMW of the NT, and of course to public health in global terms, due to the possibility of these elements becoming associated with the surface waters of the region (rivers and reservoirs), common groundwater and even the quality of the air we breathe. It will also be very important to make an inventory of all the potential sources of pollution, both local and regional, including industrial parks, with a characterization of the various existing industries, car workshops, namely those that paint and wash vehicles automatically, and places of the waste incineration, among others.

Acknowledgements

FCT – Foundation for Science and Technology supported this work with Portuguese funds within the GeoBioTec Centre (Project UIDB/04035/2020). We would like to thank Termalistur, Termas de São Pedro do Sul, E.M., S.A., Portugal, for the financial support for this publication.

Conflict of interest

The data sets generated during and/or analyzed during the current study are available from the corresponding author on request after the publication of work.

Author details

L.M. Ferreira-Gomes[1,2]*, Ana I.S. Tavares Jorge[3], Liliana F. Gonçalves[3] and S.M. Soares Almeida[4]

1 Beira Interior University, Covilhã, Portugal

2 Centre GeoBioTec (UA), Aveiro, Portugal

3 Termalistur, Termas de São Pedro do Sul, S.A., Portugal

4 Department of Civil Engineering and Architecture, UBI, Covilhã, Portugal

*Address all correspondence to: lmfg@ubi.pt

References

[1] Ferreira-Gomes LM, Jorge A, Rodrigues L. Good practices of quality control in the area surrounding of natural mineral water abstraction of São Pedro do Sul medical spa (Portugal). IOP Conf. Ser.: Earth Environ. Sci. 2021;**906**: 012092

[2] Almeida S, Gomes LM, Oliveira A, Carreira P. Contributions for the understanding of the São Pedro do Sul geohydraulic and thermomineral system: Hydrochemistry and stable isotopes studies. Geosciences. 2022;**12**(84):2022

[3] Coelho Ferreira P, Ferreira Gomes L, Oliveira AS. Contributions for conceptual geohydraulic model of the underground hydric resources of Meda municipality. KnE Engineering. 2020; **5**(5):241-255

[4] Ferreira-Gomes LM. Relatório sobre as Obras da Nascente Tradicional de SPS e Qualidade do seu Recurso. UBI: Covilhã (Original Internal Report); 2015. p. 72

[5] Ferreira-Gomes LM, Albuquerque FJA, Fresco H. Protection areas of São Pedro do Sul Spa. Portugal. Engineering Geology. 2001;**60**(1–4): 341-349

[6] Amorim Ferreira H. O Clima de Portugal. Normas climatológicas do Continente. In: Açores e Madeira correspondentes a 1931–1960. 2ª ed. Lisboa: SMN; 1970. p. 207

[7] Bonnet M, Landreau A. Mecanism d'acquisition de la qualité chimique d'une eau souterraine—Etude experimental in situ. Orléans: SGN 661 HYD-BRGM; 1979

[8] Nordemmann LMM, Bertoli JLR, Cunnha AC, Palombo CR. Análise química preliminar das águas da chuva de Cubatão. Impactos ambientais. In: Simp. BHR H. 5. ABRH, Blumenau. 1993

[9] Schoeller H. Eaux souterraines: hydrologie, dynamique et chimique. Paris: Masson & Cie; 1962. p. 642

[10] Hu GP, Balasubramanian R, Wu CD. Chemical characterization of rainwater at Singapore. Chemosphere. 2003;**51**(8): 747-755

[11] Meera V, Ahammed MM. Water quality of rooftop rainwater harvesting systems: A review. Journal of Water Supply: Research and Technology-Aqua. 2006;**55**(4):257-268

[12] Selala MS, Thenga H, Jewitt GPW, Chaplot V. Comparison of the chemical quality of rainwater harvested from roof and surface run-off systems. Water SA. 2018;**44**(2):223-231

[13] Emsley J. Nature's Building Blocks: An A-Z Guide to the Elements. Oxford: University Press; 2003. p. 538

[14] Dec.-Lei n.° 152/2017, 1.ª série—N° 235, 7 de dezembro. Water quality regime for human consumption. D.R., Lisboa, 2017. pp. 6555-6576

[15] Fetter CW. Applied Hydrogeology. 3rd ed. New York: Macmilan; College Publishing Company; 1994. p. 691

Chapter 8

Pollution Potential of Natural Sulphurous Groundwater from the Use of Geosynthetics in Underground Works Near Mineral Water Abstractions for Medical Spas

Luís M. Ferreira-Gomes and Francisco Riscado dos Santos

Abstract

Geosynthetics are used in underground works, namely geotextiles for drainage and geomembranes for waterproofing. Because some groundwaters are aggressive to the materials they contact, as is the case of the sulphurous waters used in medical spas, the question arose as to whether those materials might be degraded and, in the process, contaminate the natural groundwaters. The appearance of unusual chemical elements in the waters of the medical spa is enough to be considered contaminated and therefore leads to the closure of those establishments. Once the question was raised, an experimental plan was developed to acquire some knowledge about the situation. Thus, in this chapter, after an introduction on the importance of the subject, and a brief survey on the state of the art, the geosynthetic materials studied are presented in detail, as well as the chemical composition of virgin groundwater involved in the process. The methodology implemented is presented, and the results are shown and discussed. Finally, the main conclusions on the evolution of the physical and mechanical parameters of the geosynthetics over time (8 months of study) are presented, with special focus on the chemical changes in groundwater quality when geosynthetic materials are used in contact with them.

Keywords: sulphurous groundwater, natural mineral water, geosynthetics durability, groundwater contamination, medical spa

1. Introduction

The groundwaters of the sulphurous type are very special by heaving characteristics that give therapeutic properties. For this reason, they are frequently classified as natural mineral waters with qualifications for thermalism. Thermalism is the activity that is practiced in a Medical Spa, which functions as a particular hospital where its primary medication is natural mineral water, which is used according to medical prescription, as a consequence of the user's health problem, or eventually, used only as actions of well-being and relaxation.

The practices used in a medical spa are various types, being organized into two main groups: i) balneotherapy, with immersion techniques in the swimming pool, simple immersion bath with still water, hydromassages, shower techniques, such as Vichy massage, circular shower, and steam techniques, such as Turkish bath, hammam bath, Bertholaix, spinal steam, and foot and hand steam; and ii) ORL - Respiratory Tracts, such as nasal irrigation, aerosol, spraying and nebulization.

The diseases treated are essentially pathological rheumatic and respiratory.

Natural mineral water has a very well-defined chemical composition and keeps high stability in its chemical quality. Besides being suitable in microbiological terms, i.e., those may create health problems without pathogenic microorganisms.

To the quality and chemical stability be adequate, among several aspects, the following should be highlighted: the exploitation flow rate should never exceed the flowing torrent considered as admissible, in order not to exceed the laminar regime of the underground flow; the protection perimeter must be strictly respected, it means, the circulation of people and goods, urban occupation, construction of works, excavations, use of materials, among many others, cannot in any way compromise the rules to have complied, especially in the Immediate Protection zone of sulphurous water abstraction (the very close area around the abstraction).

In medical spas, there are often several springs of sulphurous water, which are sometimes necessary many times to drain them, which is a current situation is done using geotextiles. On the other hand, in proximity areas of the abstraction sites, it is necessary to carry out waterproofing to avoid infiltrations of surface water, having to use flow barriers, as geosynthetics geomembranes that have these functions may come into contact with sulphurous waters.

It must be emphasized that it is common knowledge that sulphurous waters are very aggressive to some materials because they quickly accelerate their degradation, even leading to the need to replace them, with significant immediate economic loss and other consequences.

This way face the presented consequences, with the tendency to use the geosynthetics in various works associated with sulphurous waters, the question raised about the durability of those materials, not only because it is essential to know about the achievement of its functions over time, but also because there is a need to know if those possible changes in the materials can put in danger the quality of the groundwater, which is required to be of very high stability since it is used as a medication, as already mentioned. In case the sulphurous water includes chemical elements that are not part of its normal quality, it will be considered contaminated, which with high probability can carry to the closure of the respective medical spa.

Studies about the durability of geosynthetics have been frequent, mainly researching the effect of ultraviolet rays, temperatures, leachate of sanitary landfill, acid, and basic solutions, among others [1–7].

Studies on the durability of geosynthetics when immersed in situations with sulphurous groundwater have been carried out by the working group of the authors of this chapter, deserving reference to the works carried out based on permeability tests of geotextiles [8], on resistance tests (punching and tensile traction) in a geomembrane and in various geotextiles [9, 10], in tearing strength tests in geotextiles [11], and in several joint situations for various geosynthetics with some conclusions on potential changes in sulphurous groundwater [12]. Studies about groundwater quality when in contact with geosynthetics are rare. However, there are some exceptions, such as studies with geotextiles to be incorporated into permeable pavements to show that their use improves the water quality that infiltrates the surface of permeable pavements [13].

2. Studied materials and methodology

The geosynthetics used in this investigation are three geotextiles and one geomembrane (**Figure 1**). The geotextiles are non-woven and use three different types: one geotextile with mechanical bonding (Needled - A), a geotextile with chemical bonding (Chemical - Q) and a geotextile with thermal bonding (Thermal - T). Geotextiles

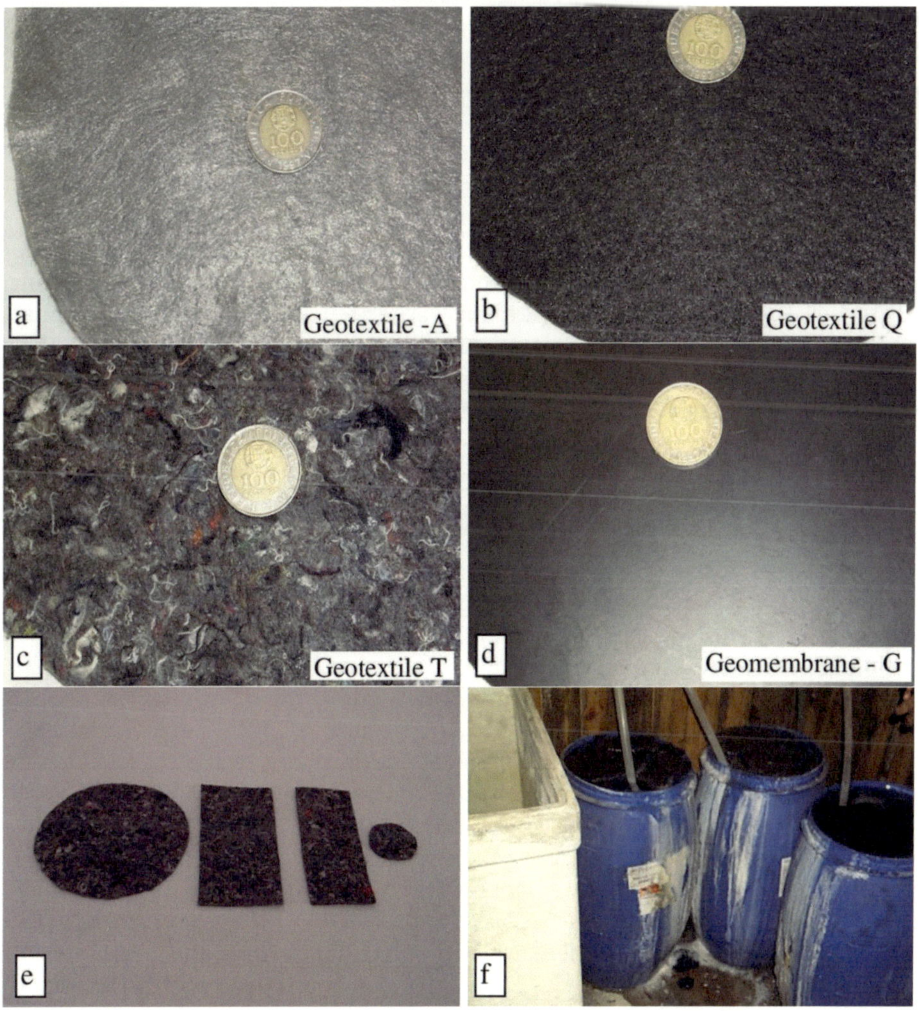

Figure 1.
Photographs of samples of studied geosynthetics, in various situations: a) to d) intact samples; e) examples of specimens subjected to punching, tensile, tearing and permeability tests; f) specimens in reservoirs with running natural sulphurous water (in continuous renovation), with evidence of the development of whitish biojelly, close to the sulphurous groundwater abstraction.

A and Q are made of polyester and polypropylene, respectively, while the geotextile T is made of approximately 15% polypropylene and the balance of recyclable materials from the textile industry polyester, cotton and wool. The geomembrane (G) is high-density polyethene (HDPE).

The mineral water used in this study comes from a medical spa with natural groundwater of the sulphurous type.

The geosynthetics were cut into specimens (of suitable size for the tests to be carried out), which are organized into three groups, according to the following: i) specimens submerged in reservoirs with sulphurous water without circulation (stagnant water – saw), at a constant temperature of 20°C, in the laboratory; ii) specimens submerged in reservoirs with running sulphurous water, in continuous renovation (running water – rw), at a natural temperature of 37°C, close to the spring of the Medical Spa, inside a small house dully adapted to the purpose; iii) specimens adequately stored, protected from any action that might alter their quality, i.e., a shadowy place, protected from light, humidity, and dust, at a temperature on about 20°C.

The reason for studying specimens immersed in reservoirs with standing sulphurous water is related to the need to acquire more sensibility about what can happen in terms of chemical alteration of the water after being in contact with geosynthetics. In nature, there is always continuous movement of groundwater within what is expected in a confined aquifer system associated with artesian springs. Nevertheless, on the other hand, it is more difficult to find chemical changes in the water due to the continuous renewal in running water. In any case, it is admitted that the reactive power of sulphurous water, in continuous renewal, may be much higher than that of stagnant sulphurous water, the reason why the study was carried out under these conditions.

The specimens i) and ii) were tested over 6 months and in the eighth month. The specimens iii) were only tested at the beginning, middle, and end of the works so that their results could be used as references. The physical and mechanical tests performed to the geosynthetics were mass per unit area [14], permeability tests [15], static puncture test, California bearing ratio (CBR) type [16], wide-width tensile test [17] and tearing strength test [18].

Regarding natural mineral water (sulphurous water), classical physical-chemical analyses were performed in the situations [19]: i) young water (yw), and ii) aged water (aw) after contact with the geosynthetic specimens, and in situation iii) aged without contact with any specimen, to the results as a reference.

Sulphurous waters are very common in a large group of Portuguese medical spas. As a rule, these waters in nature have pH > 7, SiO_2 > 10% concerning total mineralization, F^- > 5 mg/L, Sulphur in reduced, unstable forms (HS^- e $S_2 O_3^{2-}$), and the presence of HCO_{3-} and Na^+ as the dominant anion and cation, respectively. Other important elements, such as Cl^-, SO_4^{2-}, CO_3^{2-} and $H_3SiO_4^-$ in the anions, and Ca^{2+}, K^+, Mg^{2+}, and Li^+ in the cations. Some gases such as CO_2 and ^{222}Rn are also present, and some trace chemical elements such as Mn^{2+}, Br^{2-}, B_3O_3, W, Zn, Sb, and Mo.

This type of sulphurous water, if not contaminated by external agents, has excellent stability over time, as shown by several authors [20, 21]. To have a particular notion of the trace of chemical elements of the water understudy, the results of a detailed physical-chemical analysis [22] of the same groundwater abstraction that serves as the basis for this study are presented.

It should be noted that these waters aim to develop a whitish gelatinous residue (biojelly) which is typical of this type of sulphurous water (**Figure 1f**).

Scanning Electron Microscopy and Elemental Analysis techniques were also used in the characterization of the geosynthetic materials, before and after the contact with sulphurous water, as well as in the characterization of the biojelly. The equipment used is from the Laboratory of the Optics Centre of the University of Beira Interior. The scanning electron microscope is Hitachi, model S-2700, resolution 40 Å. The detector for elemental analysis, through energy dispersive RX, is a Rontec.

3. Results and interpretation

3.1 Physical and mechanical tests on geosynthetics

Physical and mechanical tests were essential, especially to evaluate the durability of these materials when in contact with this type of sulphurous groundwater. In addition, if their characteristics evolve, some chemical elements of these materials will probably become part of the chemical composition of the waters where they are installed, namely in new trace chemical elements.

About the physical and mechanical results, the characteristics of geosynthetics over the 8 months of research were presented in previous works [8–12]. It was shown that the geosynthetics immersed in these environments have evolution, namely their strength, in most situations studied, decreases over time.

Table 1 summarizes the main trends over time regarding the parameters obtained in the various tests. It is mentioned that, regarding strength parameters, although it was the thermal geotextile that suffered the highest rates of decrease (−3.99%/month) of Ts, in T-rw), there are situations in those they even gained strength (+0.95%/month of α_f, in T − sw). Other increases also occurred in the geotextiles A (e.g. +4.01%/month of Ts and + 0.71%/month of Fp). In the case of Geotextiles Q and Geomembranes G, the strength parameters in total always showed decreasing rates over time, with the most difficult situation in G, with −2.33%/month of α_f, in rw situations.

An interesting particularity was verified in the results of the permeability type test, verified in the geotextiles over time. In terms of velocity index (Iv), the results in rw and sw environments, over the time, are presented in **Figure 2**. The geotextiles in contact with the sulphurous water become more permeable during the first months (Iv increased), which is a consequence of the aggressiveness of the sulphurous water, assuming that may have the possibility that some chemical reactions occurred between the geosynthetics and the sulphurous water.

However, a particular situation occurred after some time, which was the inversion of the tendency until then verified, that is, the permeability decreased again, verified by the successive decrease of Iv; this inversion situation was followed by the appearance of a whitish residue that was developing, called "biojelly", which was more evident in the situation of running water, at the end of the fourth month; the biojelly, in besides to filling the geotextiles by external deposition, also blocks them internally.

A view of biojelly can be seen in **Figure 1f**, with the more evident development, in the borders of the reservoir, where the current sulphurous water that drains directly by artesian flow from the abstraction is discharged at the bottom of the container and flow on the top. The geotextiles in the case of the sw environment also developed biojelly, despite at a much lower rate than in the case of the rw environment.

Biojelly is a typical natural product of sulphurous groundwater, practically unknown by the community of professionals who apply geosynthetics in underground works. Studies on such product are practically non-existent, deserving reference the work of Calado [23] on the deep aquifer systems of this type of water, where several times he mentions the biojelly as a particularity of this type of groundwater.

Geos.	Situation	Parameter - equation	%/ month	Geos.	Situation	Parameter - equation	% / month
A	rw	$\mu_A = +0.974\,t + 155.7$	0.63	T	rw	$\mu_A = -7.345\,t + 594.0$	−1.24
	sw	$\mu_A = -0.311\,t + 154.3$	−0.20		Rw	$\mu_A = -5.208\,t + 600.5$	−0.87
	rw	$Fp = +0.010\,t + 1.4$	0.71		rw	$Fp = -0.001\,t + 0.2$	−0.50
	sw	$Fp = -0.017\,t + 1.5$	−1.13		sw	$Fp = -0.009\,t + 0.3$	−3.00
	rw	$\alpha_f = -0.004\,t + 6.4$	−0.06		rw	$\alpha_f = -0.080\,t + 2.2$	−3.64
	sw	$\alpha_f = -0.044\,t + 6.4$	−0.69		sw	$\alpha_f = +0.019\,t + 2.0$	0.95
	rw	$Ts = +7.926\,t + 197.5$	4.01		rw	$Ts = -3.003\,t + 75.3$	−3.99
	sw	$Ts = -0.0287\,t + 231.9$	−0.01		sw	$Ts = -1.210\,t + 71.2$	−1.69
Q	rw	$\mu_A = +0.404\,t + 154.5$	0.26	G	rw	$\mu_A = -1.174\,t + 988.1$	−0.12
	sw	$\mu_A = -0.255\,t + 157.4$	−0.16		sw	$\mu_A = 3.353\,t + 974.4$	0.34
	rw	$Fp = -0.011\,t + 2.2$	−0.50		rw	$Fp = -0.027\,t + 2.8$	−0.95
	sw	$Fp = -0.033\,t + 2.3$	−1.43		sw	$Fp = -0.030\,t + 2.9$	−1.05
	rw	$\alpha_f = -0.010\,t + 8.9$	−0.11		rw	$\alpha_f = -0.504\,t + 21.5$	−2.33
	sw	$\alpha_f = -0.141\,t + 9.8$	−1.44		sw	$\alpha_f = -0.285\,t + 20.6$	−1.38
	rw	$Ts = -0.047\,t + 293.2$	- 0.02				
	sw	$Ts = -1.448\,t + 304.3$	−0.48				

μ_A - mass per unit area (g/m^2), Fp - maximum plunger force (kN), α_f - tensile strength (kN/m), Ts - maximum tearing strength (N), t - time (month:1, 2 .., n).

Table 1.
Trends obtained from the results of physical and mechanical tests overtime on geosynthetics (geos.) immersed in running water (rw) and stagnant water (sw) [12].

3.2 Chemical composition of waters

The results of the physical-chemical analyses of the groundwater understudy and stored over 8 months for the situations "without contact" and "with contact" with the geosynthetics are presented in **Table 2**. The same table also presents the results of physical-chemical analysis of the same groundwater, but in the "young situation" (young water- yw), to facilitate their interpretation. The water referred to as "young situation" is understood as water collected immediately after its resurgence from the natural mineral water aquifer and without contact with the geosynthetics. In a graphic format, **Figures 3** and **4** present the results of the main physical-chemical parameters of the groundwater understudy in different situations.

Regarding the global physical-chemical parameters (**Figure 3**), the first significant singularity is that the total sulphide parameter only exists in the yw situation to annul itself when the water is aged, even without contact with any geosynthetic. Then, it refers to the pH situation, which decreases with the aging of the water, being lower in the situation of contact with geotextile T; therefore, the basic water in the young

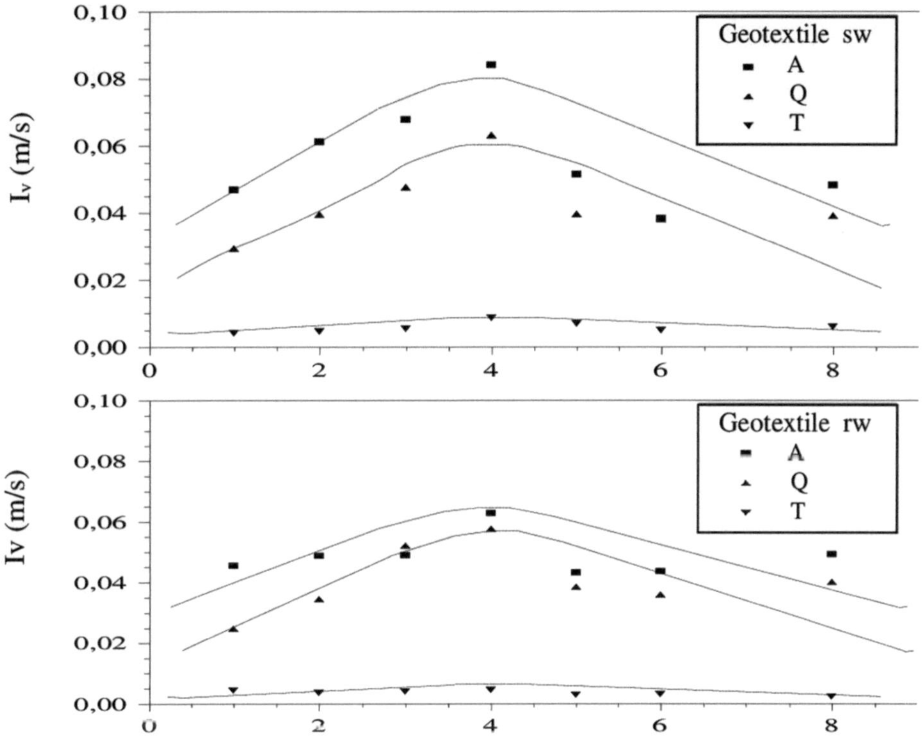

Figure 2.
Evolution of velocity index (iv) from permeability tests in geotextiles immersed in water over 8 months.

Parameter		**Young water, without contact**		**Aged water (2) (3)**				
				Without contact - aw	**After contact with geosynthetics**			
		yw (1)	**yw (2)**		**aw-A**	**aw-Q**	**aw-T**	**aw-G**
pH		8.84	8.74	7.28	7.92	7.11	6.3	8.14
Conductivity -μ S/cm		292	281	318	326	330	372	320
Total Alkalinity (in HCl 0.1 N) - mL/L		16.7	15.5	15.9	16	16.4	22.9	16.1
Dry residue (at 180°C) - mg/L		226	211	249	246	255	340	238
Total Hardness (p.p.10^5 $CaCO_3$)		0.92	0.90	1.0	1.3	1.5	3.2	1.1
Total sulphide (in I_2 0.01 N) - mL/L		11.2	11.1	<1.2	<1.2	<1.2	<1.2	<1.2
Silica (SiO_2) - mg/L		53.2	52.0	58.0	56.0	56.0	62.0	57.0
Total mineralization - mg /L		264	254	287	291	294	348	287
	Sodium (Na^+)	67.1	63.0	70	71	71	77	70
	Calcium (Ca^{2+})	3.1	3.4	3.9	4.9	5.5	9.2	4
Cations	Potassium (K^+)	1.9	1.9	1.9	2.2	1.9	5.4	1.9
(mg/L)	Magnesium (Mg^{2+})	0.12	0.17	0.15	0.18	0.21	2.1	0.16
	Lithium (Li^+)	0.30	0.27	0.3	0.29	0.31	0.3	0.3

Parameter		Young water, without contact		Aged water (2) (3)				
				Without contact - aw	After contact with geosynthetics			
		yw (1)	yw (2)		aw-A	aw-Q	aw-T	aw-G
	Ammonium (NH_4^+)	0.04	0.17	n.d.	n.d.	n.d.	0.08	n.d.
	Bicarbonate (HCO_3^-)	83.0	82.7	97.0	97.8	100	140	98.4
	Chloride (Cl^-)	25.2	22.1	26.8	26.8	27.4	31.3	26.6
	Sulphate (SO_4^{2-})	6.9	7.6	13.8	14.5	16.3	<0.3	13.8
Anions	Fluoride (F^-)	15.0	14.0	14.7	14.9	14.9	20.3	14.8
(mg/L)	Nitrate (NO_3^-)	< 0.10	< 0.3	<0.3	2.6	<0.3	<0.3	0.4
	Nitrite (NO_2^-)	< 0.002	< 0.005	<0.005	<0.005	<0.005	<0.005	<0.005
	Hydrogensulphide (HS^-)	1.8	1.8	<0.2	<0.2	<0.2	<0.2	<0.2
	Phosphate ($H_2PO_4^-$)	58	n.d.	n.d.	n.d.	n.d.	n.d.	n.d.
	Aluminium (Al_3^+)	< 20	n.d.	n.d.	n.d.	n.d.	n.d.	n.d.
	Manganese (Mn^{2+})	13	n.d.	n.d.	n.d.	n.d.	n.d.	n.d.
	Bromide (Br^{2-})	264	n.d.	n.d.	n.d.	n.d.	n.d.	n.d.
	Boron (B_3O_3)	104	n.d.	n.d.	n.d.	n.d.	n.d.	n.d.
	Beryllium (Be)	0.8	n.d.	n.d.	n.d.	n.d.	n.d.	n.d.
	Lead (Pb)	6	n.d.	n.d.	n.d.	n.d.	n.d.	n.d.
	Silver (Ag)	< 0.8	n.d.	n.d.	n.d.	n.d.	n.d.	n.d.
	Cadmium (Cd)	0.3	n.d.	n.d.	n.d.	n.d.	n.d.	n.d.
	Vanadium (V)	< 3	n.d.	n.d.	n.d.	n.d.	n.d.	n.d.
	Yttrium (Y)	< 0.1	n.d.	n.d.	n.d.	n.d.	n.d.	n.d.
Trace	Tin (Sn)	< 3	n.d.	n.d.	n.d.	n.d.	n.d.	n.d.
chemical	Chromium (Cr)	< 2	n.d.	n.d.	n.d.	n.d.	n.d.	n.d.
elements	Iron (Fe^{2+})	< 25	< 30	< 30	90	50	100	<.30
	Barium (Ba^{2+})	< 2	n.d.	n.d.	n.d.	n.d.	n.d.	n.d.
(mg/L)	Iodide (I^-)	< 1	n.d.	n.d.	n.d.	n.d.	n.d.	n.d.
x 10^{-3}	Arsenic (As_2O_3)	29	n.d.	n.d.	n.d.	n.d.	n.d.	n.d.
	Tungsten (W)	28	n.d.	n.d.	n.d.	n.d.	n.d.	n.d.
	Copper (Cu)	< 1	n.d.	n.d.	n.d.	n.d.	n.d.	n.d.
	Zinc (Zn)	13	n.d.	n.d.	n.d.	n.d.	n.d.	n.d.
	Antimonium (Sb)	8	n.d.	n.d.	n.d.	n.d.	n.d.	n.d.
	Nickel (Ni)	< 3	n.d.	n.d.	n.d.	n.d.	n.d.	n.d.
	Cobalt (Co)	< 1	n.d.	n.d.	n.d.	n.d.	n.d.	n.d.
	Niobium (Nb)	< 0.5	n.d.	n.d.	n.d.	n.d.	n.d.	n.d.
	Molybdenum (Mo)	6	n.d.	n.d.	n.d.	n.d.	n.d.	n.d.

(1) complete physical-chemical analysis carried out by the IGM Laboratory [22]. (2) summary physical-chemical analysis carried out by the IST Laboratory; (3) analyses performed after 8 months of aging; n.d. – parameter not determined.

Table 2.
The physical-chemical analyses of the sulphurous water studied, before (young water - yw), and after contact (aged water - aw) with the geosynthetics.

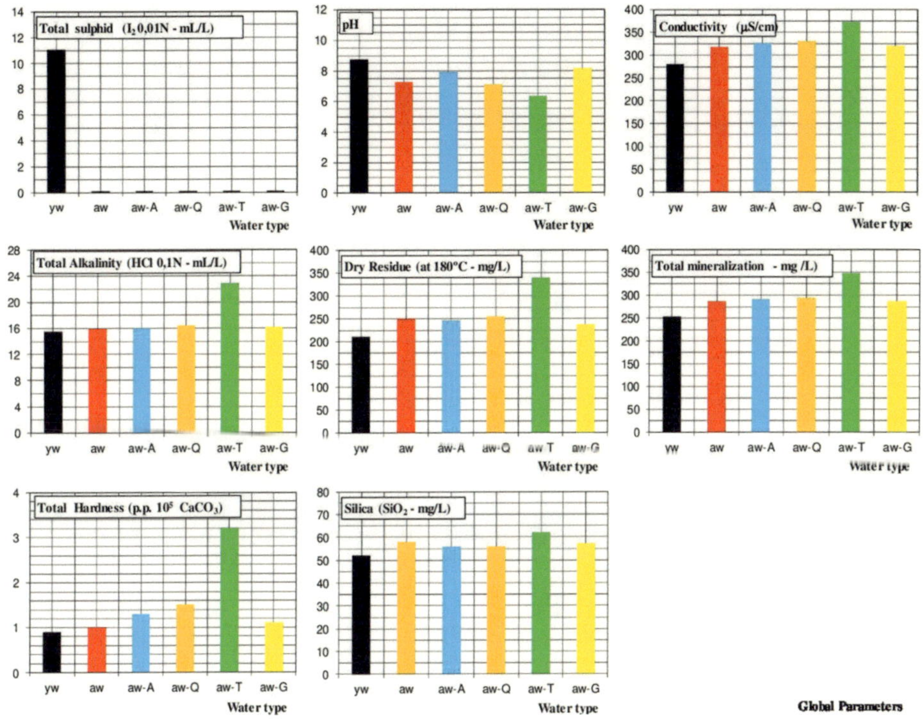

Figure 3.
Results of the global physical-chemical parameters of the groundwater understudy, in different situations: i) as young water without contact with the geosynthetics (yw); ii) as aged water without contact (aw); iii) in aged water situation with contact of the geosynthetics, needled (aw-a), chemical (aw-Q), thermal aw-T and geomembrane (aw-G).

situation changes to acidic or less basic in the aged situations. Following refers to the fact that the parameters total alkalinity and total hardness do not change much, except for aw-T, where these parameters are much higher in aged water than in young water. Finally, we emphasize the situations of conductivity, dry residue, total mineralization, and silica, in which these parameters are always higher in the aged water, including the water that has no contact with the geosynthetics.

Regarding the cations (**Figure 4a**), in global terms, there is an increment in the various parameters on aged water, especially water associated with geotextile T (aw-T), with significant increases concerning young water, in magnesium, calcium and potassium. Note the particularity of aged water "without contact" and "with contact" with other geosynthetics (A, Q, G), not having any notable differences. About anions (**Figure 4b**), there is an increment of these for most situations in aged water about the situation of young water, that is, in the following parameters: bicarbonate, chloride and fluoride; there is the exception and singularity that in aw-T, sulphate is annulled. Still, on cations (**Table 2**), the positive fact is emphasized that, in any of the aged waters, nitrites were never detected, and the case of nitrates were only detected in aw-A and aw-G.

Concerning trace chemical elements, the only element investigated was the Iron, and it looks that the aged water does not gain this element without contact, nor by the aged water that is in contact with the geomembrane; on the other hand, the increase of this same element in the water in contact with the various geotextiles, namely in aw-T and aw-A, is very significant.

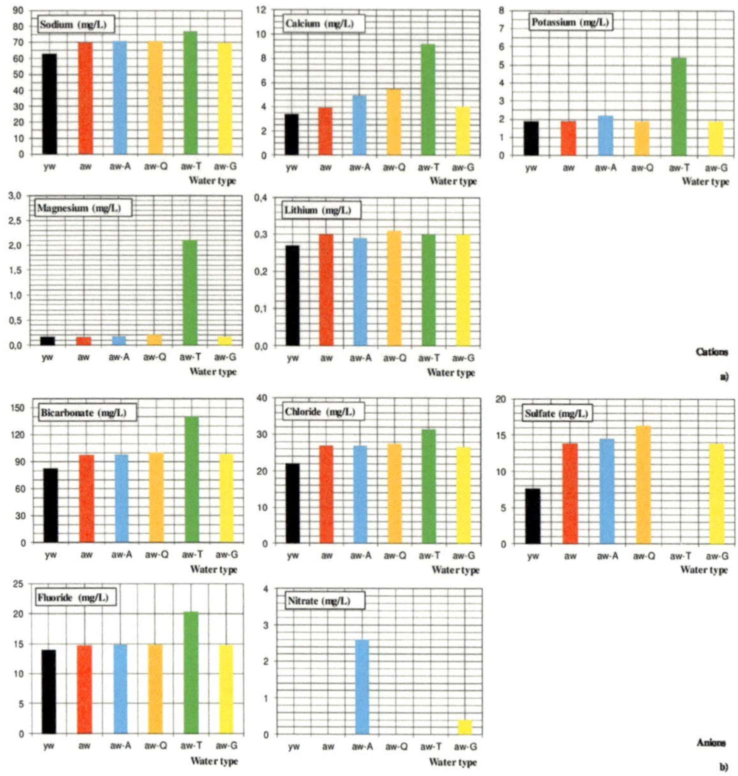

Figure 4.
Results of the main ions of the groundwater understudy, in different situations: i) as young water without contact with the geosynthetics (yw); ii) as aged water without contact (aw); iii) in aged water situation with contact of the geosynthetics, needled (aw-a), chemical (aw-Q), thermal aw-T and geomembrane (aw-G).

The water that has the most modifications when aged is the one that is in contact with the thermal geotextile (aw-T). From the analysis of the water classifications in the various situations (**Table 3**), the aged water either "with contact" and "without contact" with the geosynthetics loses in the classification, about the young water, the term "sulphurous". A situation that still needs to be highlighted is that water with thermal geotextile changes in its classification the term from "alkaline reaction" to "acid reaction".

When performing the Piper diagram for all analyses (**Figure 5**), it is verified that the aged waters remain in the same group I, as the young water, that is, they are all Sodium bicarbonate type. However, the aged water next to the type T geosynthetic (aw-T) stands out.

3.3 Scanning electron microscopy and elemental analysis

To analyse the much detail as possible the relation between the biojelly and the geosynthetics, scanning electron microscope images were taken of all geosynthetics before and after contact with sulphurous water. Examples are presented in **Figure 6** for the cases of geotextile A and geomembrane. Comparing the images of the various geosynthetics, before and after contact with sulphurous water, we found that the biojelly is fixed in the internal structure of the geotextiles, and more clearly in the geotextiles, A and T. A completely different situation is verified in geomembranes, where the evidence of biojelly is only superficial.

Sample	Water type – Classification
yw	Weakly mineralized water, sulphurous, fluoridated, with alkaline reaction and soft
aw	Weakly mineralized water, fluoridated, with alkaline reaction and soft
aw-A	Weakly mineralized water, fluoridated, with alkaline reaction and soft
aw-Q	Weakly mineralized water, fluoridated, with alkaline reaction and soft
aw-T	Weakly mineralized water, fluoridated, with acid reaction and soft
aw-G	Weakly mineralized water, fluoridated, with alkaline reaction and soft

Table 3.
Chemical classifications of the groundwater understudy in different situations: Without contact with geosynthetics, in (yw), and aged without contact with geosynthetics (aw), and aged in contact with the geosynthetics: Needled (aw-a), chemical (aw-Q), thermal (aw-T) and geomembrane (aw-G).

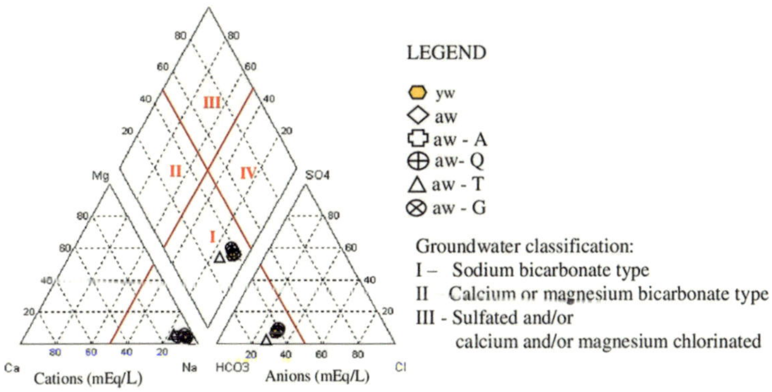

Figure 5.
Piper diagram from the physical-chemical analyses of the groundwater understudy, before (yw) and after (aw) of the contact with the geosynthetics immersed during 8 months.

Figure 6.
Examples of geosynthetic images (A and G) obtained in the electronic scanning microscope with a 200 x zoom, before (b) and after (a), from being in contact with the sulphurous groundwater.

Regarding the elemental analyses, it was not possible to obtain results for the geotextile Q and geomembrane because the analysis technique does not result in materials with those characteristics. However, it was then possible to perform the X-ray Energy Dispersive Spectra for geotextiles A and T and the biojelly situation sampled in dense and less dense areas to enable more consistent analysis of results. Typical

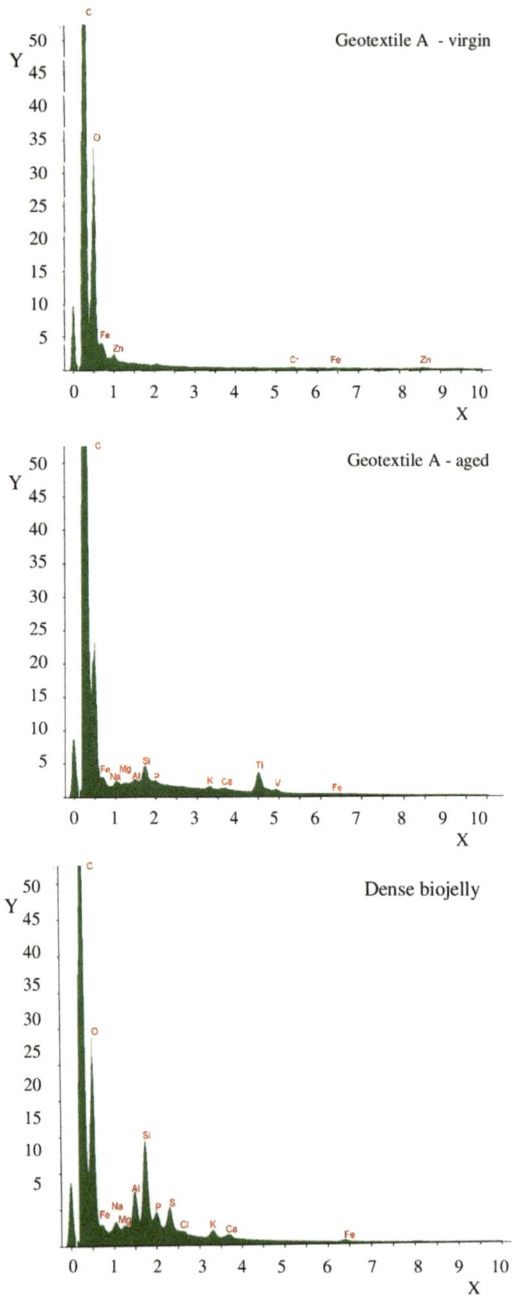

Figure 7.
Energy dispersive X-ray spectra on elementary analysis of geotextiles a in virgin and aged situations, in contact with sulphurous water, and still only of the biojelly in a high-density zone (X - energy in keV, Y - count x10³).

Sample (*)		**Main elements**	**Vestigial elements**
Geotextile A	virgin	C – 72% O – 26%	Cr, Fe, Zn
	aged	C – 95% Ti – 2%	Na, Mg, Al, Si, P, K, Ca, V, Fe
Geotextile T	virgin	C – 96.2%	Na, Mg, Al, Si, P, S, K, Ca, Ti, Fe
	aged	C – 97%	Na, Al, Si, P, S, Cl, K, Ca, Ti, Fe
Biojelly from the running water in contact with the geosynthetics	dense	C – 63% O – 28% Si – 2.6% P – 1.5% S – 1.7%	Na, Mg, Al, Cl, K, Ca, Fe
	not much dense	C – 68% O – 22% Si – 2% P – 1.6% S – 3%	Na, Mg, Al, Cl, K, Ca, Fe

(*)*it was not possible to apply this technique to the geotextile Q and the Geomembrane.*

Table 4.
Elemental analysis results from X-ray energy dispersive spectra of virgin and aged geosynthetics and biojelly samples.

spectra for the case of geotextile A before and after contact with sulphurous water, as well as the biojelly, are shown in **Figure 7** as examples of the results achieved.

Table 4 presents the global results of the various elements found in the various materials. It must be noted that carbon is always the most preponderant element both in the geosynthetics and in the biojelly itself. It is interesting to check that oxygen occurs in a significant percentage in virgin geotextile A and the biojelly and does not appear in aged geotextiles. Emphasizing that some trace elements such as Fe, Zn, Cr, Ti, Al are present in the geotextiles, those transferred to the sulphurous water could be detrimental to its quality and public health. A singularity was that geotextile A appears with Ti with some significance in the aged situation.

Finally, we mention the situation of biojelly, which is composed of carbon and oxygen, is then, with some significances, S, Si, and even P.

4. Conclusions

After the question raised: "*Will or will not the use of geosynthetics in contact with sulphurous groundwater potentiate the contamination of those waters in the aquifer system?*" a study was carried out into three main dominations: i) study of the evolution over time, of the physical and mechanical properties of geosynthetics in contact with sulphurous water; ii) study of the chemical composition of sulphurous water before being in contact and after being aged, with or without contact with geosynthetics; iii) identification of advanced studies and characterization of geosynthetics and biojelly, as complementary to the previous domains, in order to clarify the interpretation of the results.

The materials used in the study were three non-woven geotextiles, being geotextile with mechanical bonding (Needled-A), one geotextile with chemical bonding (Chemical-Q) and another geotextile with thermal bonding (Thermal-T) and also a geomembrane (G). Almost all geosynthetics are composed of synthetic fibres, with the particularity that only the geotextile T has in its composition recyclable fibres from the textile industry, some of the natural, cotton and wool, as presented in the methodology section. The biojelly was a natural product involuntarily involved in the process, constituting a whitish creamy paste, and which arose naturally from the water, associating itself with the geosynthetics, not having initially been foreseen, for being an unknown possibility. In terms of methodology, among several details, we emphasized the fact that studies were carried out in two main methods: the case of geosynthetics always in contact with rejuvenated sulphurous water (rw) and the case of geosynthetics immersed in the same water, the situation of stagnant water (sw).

Therefore, from all studies carried out, there is to re-enforce the appearance of biojelly, whitish cream (**Figure 1f**), which is evident at the end of the fourth month of investigation, with a large quantity in the rw situation. This particularity allowed to obtain more knowledge about the use of geotextiles, especially in the domain of their filling but made it challenging to analyse the results concerning the primary goal of the research, as the inclusion of biojelly in these materials interferes with the physical and mechanical evolution of some geotextiles, over time. The biojelly, very plastic paste, sticks more superficially in the geomembrane and the geotextile Q. Inside the structure of the geotextiles A and T. These situations were more clearly identified with images of these materials, virgin and aged with the sulphurous water, from the electronic scanning microscope (**Figure 6**). The biojelly, from X-ray Dispersive Energy spectra studies (**Figure 7**, **Table 4**), is has been found that it is composed essentially by C and O, those together have on about 91%, to those are added on about 2.4% S, 2.3% Si, 1.6% P, and the following trace elements: Na, Mg, Al, Cl, K, Ca, Fe.

From the studies with physical tests carried out over time, within the first three months, aggressive evident action of the water in the geotextiles was found, making them more porous and permeable, a situation particularly well clarified in the results of the permeability tests, from the evolution of Iv over the time (**Figure 2**), more evident in geotextile A, which is the most permeable, then Q and finally T. At the end of the third month, approximately, the effect of the biojelly started to be noted, and, as it increased, the permeability decreased, reflecting in the lowering of the Iv index. However, degradation of the mechanical characteristics of the geosynthetics continued to occur. It is noted that the geomembrane was not naturally subjected to the permeability test, as it is an impermeable material.

Regarding the mechanical properties of geosynthetics in contact with sulphurous water over the time, it should be noted that almost in general, they evolve towards degradation, and in particular, the various strength parameters studied (Fp - maximum plunger force, α_f - tensile strength, and Ts - maximum tearing strength), decrease over the time (**Table 1**). The geosynthetics that suffered more changes were the thermal type (T), and the minor changes were the needled type (A), and it should be noted that the first ones reached rates of about −3.99%/month of Ts in an rw environment. In the case of the geotextile A, there is a positive singularity. In an rw environment, there was positive evolution, which means, it was found increases in strength over time, with the most significant situation of +4.01%/month of Ts in an rw environment; this kind of situation is explained by the fact that in the rw environment they have developed immense biojelly, with its intertwining with synthetic fibres; it is also admitted that the appearance of the chemical element Ti in the aged geotextile A also favours this situation. In the case of the geomembrane (G), where the biojelly was permanently restricted to the surface, the strength parameters in all situations always showed decreasing strength rates over the time. The most difficult situation was in $\alpha_f = -2.33\%$/month in the rw environment. The situation close to G was shown in the geotextile Q, where the biojelly did not penetrate easily into the geotextile structure, and almost always, there showed decreases in strength.

About the sulphurous water quality over the time studied only in sw environment. Several changes exist concerning virgin sulphurous water, including aged water without contact with geosynthetics. All aged waters, due to these changes, are loose in their classification: the term "sulphurous" (**Table 3**), due to the cancelation of the "total sulphide" (**Table 2**). Besides this singularity and still, about the classification, the aged waters almost all keep the same classification, as "Weakly mineralized water, fluoridated, with alkaline reaction and soft", except for the aged water with the geotextile T, which instead of presenting an "alkaline reaction", changes to an "acid reaction".

A detailed analysis in chemical terms shows that some changes of quality occur in the water in the sw situation, mainly when in contact with the geosynthetics; from this analysis, the water that has the more significant change is that which is in contact with the geotextile T, followed by the Q, then the A, and finally the geomembrane. We emphasize the case of aged water associated with geotextile A and geomembrane as not having a significant change, and its quality is even close to aged water without contact with geosynthetics.

Finally, in global terms, it is emphasized that the geotextile A type and the geomembrane G that led to minor changes in sulphurous groundwater quality, which are indicated as the most favourable for the use of works in medical spas areas. In any case, it is understood that it must have more new similar studies to those presented in this paper, in more dilated periods, with a broader variety of geosynthetics, and with the realization of a wider variety of trace elements in the aged waters.

Acknowledgements

FCT – Foundation for Science and Technology supported this work with Portuguese funds within the GeoBioTec Centre (Project UIDB/04035/2020). We would like to thank Termalistur, Termas de São Pedro do Sul, E.M., S.A., Portugal, for the financial support for this publication.

Conflict of interest

The datasets generated during and/or analysed during the current study are available from the corresponding author on request after the publication of work.

Author details

Luís M. Ferreira-Gomes[1,2]* and Francisco Riscado dos Santos[3]

1 Beira Interior University, Covilhã, Portugal

2 Centre GeoBioTec (UA), Aveiro, Portugal

3 ADC - Águas da Covilhã, Covilhã, Portugal

*Address all correspondence to: lmfg@ubi.pt

References

[1] Koerner RM. Durability and Aging of Geosynthetics. Canada: Elsevier Applied Science; 1989. p. 332

[2] Barone FS, Costa JMA, Ciardullo L. Temperatures at the base of a municipal solid waste landfill. In: Fiftieth Canadian Geotechnical Conference, October, CGS, Ottawa. 1997. pp. 144-152

[3] Hsuan YG, Koerner RM. Antioxidant depletion lifetime in high-density polyethylene geomembranes. Journal of Geotechnical and Geoenvironmental Engineering. 1998;**124**(6):532-541

[4] Rowe RK, Sangam HP. Durability of HDPE geomembranes. Geotextiles and Geomembranes. 2002;**20**:77-95

[5] Kay D, Blond E, Mlynarek J. Geosynthetics durability: A polymer chemistry issue. In: 57th Canadian Geotechnical Conference - 5th Joint CGS/ IAH-CNC Conference, Session 4D. Quebec: CGS & IAH; 2004. p. 14

[6] Koerner GR, Hsuan YG, Koerner RM. The durability of geosynthetics. In: Sarsby RW, editor. Geosynthetics in Civil Engineering. Cambridge: CRC Press; 2007. pp. 36-64

[7] Greenwood JH, Schroeder HF, Voskamp W. Durability of geosynthetics. CUR, Building & Infrastructure, Report 243. 2012:295

[8] Riscado dos Santos F, Ferreira-Gomes LM, Figueiredo NM. Estabilidade de parâmetros hidráulicos de geotêxteis em ambientes de águas minerais sulfúreas. In: 8° Cong. Nacional de Geotecnia. Lisboa: SPG-LNEC; 2002. pp. 1837-1847

[9] Ferreira-Gomes LM, Riscado dos Santos F, Figueiredo NM. Durabilidade de geomembranas em ambientes de águas minerais sulfúreas. Geotecnia. 2005;**104**:57-72

[10] Ferreira-Gomes LM, Riscado dos Santos F, Figueiredo Nunes M. Durability based on resistance tests of geosynthetics submerged in a sulphurous environment. In: Kuwano, Koseki J, editors. Proc. 8th Int. Conf. on Geosynthetics. Yokohama: Millpress; 2006. pp. 1551-1554

[11] Ferreira-Gomes LM, Riscado dos Santos F, Figueiredo Nunes M. Tearing strength tests in geotextiles to contribute in the evaluation of its durability in sulphurous groundwater environment. In: Palmeira EM, Vidl EM, Sayão ASJ, Ehrlich M, editors. Proc. 9th Int. Conf. on Geosynthetics. Guarujá; 2010. pp. 843-846

[12] Ferreira-Gomes LM, Andrade Pais LJ, Riscado dos Santos F, Barata dos Santos PM. Stability of geosynthetics in natural sulphurous groundwater. In: Ziegler et al., editors. 10th Int. Conf. on Geosynthetics. Berlin, Germany: Deutsche Gesellschaft fur Geotechnik e. v., German Geotechnical Society; 2014. p. 8

[13] Scholz M. Water quality improvement performance of geotextiles within permeable pavement systems: A critical review. Water. 2013;**5**(2):462-479. DOI: 10.3390/w5020462

[14] EN (European Standard). 965 - Geotextiles and Geotextile-Related Products-Determination of Mass per Unit Area. European Union: European Committee for Standardization (CEN); 1995. p. 4

[15] EN-ISO (European Standard-International Organization for Standardization) 11058 - Geotextiles

and Geotextile-Related Products - Determination of Water Permeability Characteristics Normal to the Plane, without Load. European Union: European Committee for Standardization (CEN); 1998

[16] EN-ISO (European Standard-International Organization for Standardization) 12236 - Geotextiles and Geotextile-Related Products-Static Puncture Test (CBR-California Bearing Ratio Test). European Union: European Committee for Standardization (CEN); 1996

[17] EN-ISO (European Standard-International Organization for Standardization). 10319 - Geotextiles - Wide - Width Tensile Test. European Union: European Committee for Standardization (CEN); 1996

[18] ASTM (American Society for Testing and Materials). D4533 - Standard Test Method for Trapezoid Tearing Strength of Geotextiles. ASTM International; 1991

[19] Riscado dos Santos F. Estabilidade físico-mecânica e hidráulica de geossintéticos em águas minerais termais. [Master's thesis]. Portugal: University of Beira Interior; 2002. p. 108

[20] Ferreira-Gomes LM, Maia de Carvalho PE, Mendes E, Andrade Pais LJ. Hydrochemistry of water from a deep hole with objective to its legalization. In: Antunes IMHR, Almeida JPF, Albuquerque MTD, editors. VII Cong. Ibérico de Geoquímica and XVII Semana de Geoquímica. Vol. II. Castelo Branco, Portugal. 2011. pp. 607-612

[21] Coelho Ferreira PJ, Ferreira-Gomes LM, Carvalho PEM, Oliveira AS. Water quality in deep and confined aquifer systems of granite rocks - the case of sulphurous water from Longroiva, Portugal. In: Proc. of the Int. Conf. On Civil Engineering. Singapore: CI-Premier PTE LTD; 2013. pp. CE115-CE124

[22] IGM – Instituto Geológico Mineiro. Estudo Físico-Químico completo da água do Furo ACP1 das Termas de Unhais da Serra. Rel. 11/94. São Mamede de Infesta: IGM Laboratório; 1994. p. 7

[23] Calado CMA. A Ocorrência de Água Sulfúrea Alcalina no Maciço Hespérico: Quadro Hidrogeológico e Quimiogénese [PhD Thesis]. Portugal: University of Lisboa; 2001. p. 461

Section 4

Management Strategies of Water Quality

Chapter 9

Safe Drinking Water: The Need and Challenges in Developing Countries

Collins Onyebuchi Okafor, Ude Ibiam Ude, Felicia Ngozi Okoh and Blessing Osose Eromonsele

Abstract

Safe or potable water is water which quality is suitable for human consumption both for drinking or cooking purposes. Safe and potable drinking water is an inevitable factor for a robust healthy society, but clean water for drinking and domestic purpose still remains inaccessible to 1.1 million people on a global scale. The importance of potable water can never be overestimated because of its benefits of providing a number of useful services for humans and the ecosystem. Water contributes directly and indirectly to the GDP of a country when tailored properly. While the majority of the population in developing countries lives in cities, supply and access to safe drinking water in sufficient quantity and quality for all urban dwellers sits at a crossroads between environmental issues. For instance the issue of available freshwater resources, and development issues, through the implied social, economic and sanitation challenges still remain unsolved. In order to achieve quality drinking water in fostering human development, some vital issues that pose challenges must be taken into good consideration. This chapter has discussed safe drinking water in view of its benefits and challenges as it affects developing countries.

Keywords: water, quality, health, benefits, challenges

1. Introduction

Water is safe and fit for drinking as a result of been free from biological (pathogenic microorganisms), physical (debris and impurities) and chemical (toxins, carcinogens and radionuclide) contaminants or other health hazards. For water to be drinkable, it must be devoid of unpleasant tastes, odors and colors and maintained at a reasonable and acceptable limits of temperature. In order to satisfy the increasing demand for safe and potable water supply, the cleanest sources of surface water and groundwater must be preserved for such purposes. Potable water whether treated or raw (i.e untreated) should at least meet numerous physical, chemical, biological and microbiological standards [1, 2].

Safe and potable drinking water is an inevitable factor for a robust healthy society, but clean water for drinking and domestic purpose still remains inaccessible to

1.1 million people on a global scale. In developing countries, sources of drinking water ranges from surface water, groundwater (such as well water and borehole water), spring water, saline water, bottled water and harvested rainwater. The presence of opportunistic and obligate pathogens such as *Pseudomonas* spp., *Aeromonas* spp., *Klebsiella* spp., *Mycobacter* spp., *Escherichia coli, Helicobacter* spp., *Salmonella* spp. and *Legionella* spp., which may increase the health risks associated with the consumption of water, arises from these outlined sources. According to World Health Organization, the global estimate on death resulting from diarrhea as a waterborne disease exceeds 1milion annually, many of which have been linked to diseases acquired from the consumption of contaminated waters and seafood. In some countries, cases of water related diseases are inestimable because of lack or poor health data, inefficient treatment strategies and policy implementations. The challenge of maintaining potable and quality drinking water is the leading source of outbreaks of water related diseases across the globe. This calls for new approaches in the treatment of water meant for public consumption [3].

2. Benefits of safe drinking water

Potable or safe drinking water is extremely important to human life and human development because water is a "sin qua non" for almost every human activities and indispensable for sustainable development. The negative impact of unsafe water over health, national productivity, quality of life and socio- economic development are enormous. This is because water is a decisive component in human survival with respect to security, socio-economic activities, which include transportation, agriculture, production, domestic and foreign consumption and maintenance of healthy ecosystems [4, 5].

Water is a lifeline of the biosphere and planetary life that requires adequacy, uniform distribution, reliability, equal accessibility, clean, safe and acceptability [6]. When safe and clean water is uniformly distributed and the demand for it remains lower than the available supply, it rapidly fosters exponential growth that support social and economic development, ensures national security, promotes and maintains healthy ecosystems across different spheres of life in many regions of the world [4, 7]. When the quality of water is satisfactory, it becomes a fundamental indicator of good public health, societal well-being and a crucial element for rapid development of a country. On the other hand when water is contaminated, it becomes a vehicle of diseases and a potential to threat to human, thus affecting individual productive rate [8, 9].

According to the WHO [10], about 1.1 billion people consume unsafe water across the globe contributing to about 1.7 million global death annually (approximately 3.1%) with a potential annual disability burden of 54.2 million (~3.7%). This calls for the provision of safe water and basic water related sanitation and hygiene practices. In developing countries, the benefit of safe and potable water is not sufficiently documented as a result of poor or nonexistent government interest for water related issues and low level of investment in water treatment and distribution infrastructures. It may interest you to know that from little information gathered in these countries, water contributes immensely to the little leap experienced by their staggering economic development. In developing continents like Africa and Southern America, water alone contributes significantly to their economy with a very high benefit-cost ratio of more than 5:1 due to its intertwined connection with industrialization, agriculture/food and energy production, whereas in some other countries, it may be as high as 7:0 but with a minimal recorded of 2:0 across the globe. This mean that if developing countries pay maximum attention to ensuring adequate management of water resources, there will be a resultant gain in economy ranging from 5 to 30 USD per 1USD investment beside other technical, environmental and political gains [9, 11]. Apart contributing to economic development, the water sector is extremely important in the development of other aspect of national growth because of its interconnected relevance with other developmental sectors such as the household, health, agriculture and food, energy and power, industry and manufacturing, environmental, educational, legal and political at local, national, regional and international levels [12].

When sources of potable water such as rivers, aquifers, and underground waters rivers become heavily polluted, downstream regions usually experiences loss in economic growth ranging between 0.8 and 2.0 percent of economic growth. This is because economic growth and water pollution are intrinsically related. Inasmuch as quality water improves national economy, declining water quality can impact the economy in various ways across the health sector where labour productivity can be adversely affected, agricultural sector in terms reduction in quality and quantity of food produced, tourism, real estate, aquaculture/fisheries and other sectors which rely on environmental quality and ecosystem services. Irrespective of these well-known impacts, estimating the impact of water quality on economic activity can be very slippery. Therefore is a need to fill this gap by using a run-of-the-mill practical approach in modern environmental economics and new data on economic activity and water quality. Because water is always needed for life, health, and economic production, the impurities generated by predicate polluters may affect succeeding users [13]. Undoubtedly, access to potable water has series of direct and indirect benefits with respect to health, education, poverty and environment as there exist a connecting nodes between water and sustainable development, far beyond its social, economic and environmental dimensions because safe water plays a significant role in addressing the developmental challenges, such as human health, food and energy security, urbanization and industrial growth, as well as climate changes [14].

In the past two decades, about 2.3 billion people have gained access to an improved drinking water indicating a monumental gain, but there's a lot to be done however, this success is somewhat teetering as many of developing countries, especially the poor ones are still battling to get access to potable as research has shown that the most of the people without access to safe water are from developing nations implying many people in the developing world, especially Africa, still depend on unsafe water sources for daily water need and affected by chronic water problems and water-borne diseases [15–18]. Notwithstanding, safe (potable) water remain the main developmental pillars of national development that should be embraced by the developing countries.

3. Challenges of safe drinking water in developing countries

Several factors are germane to ensuring the availability of potable water with respect to the increasing demand by the growing population. The factors that are key interest in this chapter include availability and scarcity, contamination, portability, accessibility and affordability, sustainability and climate change.

3.1 Contamination

Contamination of natural waters represents one of the main risks to public health, a fact that is directly related to the discharge of untreated domestic, hospital, and industrial effluents, which cause contamination of aquatic bodies through pathogenic microorganisms such as bacteria, viruses, protozoa, and helminth eggs that are contained in them [18]. Bacterial contamination of water resources coupled with other contaminants such as nitrates, metals, trace quantities of toxic materials and salts are the greatest threat to safe drinking water. The presence of opportunistic pathogens such as *Pseudomonas* spp., *Aeromonas* spp., *Klebsiella* spp., *Mycobacter* spp., *E. coli, Helicobacter* spp., *Salmonella* spp. and *Legionella* spp. that may increase the health risks associated with the consumption of water arises from these sources and should be checkmated [3, 19–22].

Millions of people die due to water-related diseases such as cholera, diarrhea (mainly), malaria, dengue fever, colitis and so on. On a global estimate, more than 25,000 people (comprising about 5000 children) die daily from water related diseases of which most of these diseases can easily prevented if safe drinking water tops the scale of preference among leaders across the world. In most developing countries, waterborne diarrhea and other related water diseases kill about 1.8 million children per year representing the leading cause of death globally. Indicators of fecal contaminated water (i.e coliform bacteria) are ingested by almost 1.8 billion people through feacally contaminated water. In order to achieve the goal of making water safe for drinking, removal of contaminants must be a regular normal. Access to safe water entails the reduction of water-related diseases and serves as an opportunity for improved health because of it potential reduction of the outbreak of health hazards [9].

More than 5 million people die yearly from water associated disease. More than 50 percent these waterborne diseases are of microbial origin, having cholera with highest prevalence. The source of microorganisms in water sources is usually from waste water discharge into fresh waters and coastal waters. The contamination of water bodies is a regular muddle across the globe. These contaminations may be environmental or anthropogenic in nature. Higher levels of contaminants in drinking water are seldom to cause acute health effects. Of course it depends on individual susceptibility and mode of contact with the body. The microbiological examination of water is used naturally and scientifically worldwide to monitor and control the quality and safety of drinking waters [23, 24].

Some contaminants can be easily identified by assessing color, odor, turbidity and the taste of the water. However, most cannot be easily detected and require testing to reveal whether water is contaminated or not. Thus, the contaminants may result in unappealing taste or odor and staining as well as health effects. Color of the drinking water is a physical characteristic that cannot be noticed unless it is one of high concentration [25].

Groundwater contamination may occur as a result of the types and concentrations of natural contaminates. The nature of these contaminates depend on the type of the topographical materials along the flow path of the groundwater and the quality of the recharge water. For instance groundwater flowing via sedimentary rocks and soils may pick up a wide range of chemical contaminants (magnesium, calcium, and chloride, arsenate, fluoride, nitrate, and iron) in concentrations exceeding the recommended level becomes unacceptable for water meant for drinking and domestic purposes. Although, the impact of natural contamination of groundwater depends on the types and concentrations of the contaminants, when they occur in water at unacceptable levels will definitely render the water unfit for drinking [26]. Man-made water contaminants arise from by-products of household, industry and agricultural practices. These pollutants include heavy metals such mercury, copper, chromium, lead, and hazardous chemicals, dyes and compounds like insecticides and fertilizers and they represent a significant group of groundwater contaminants [25].

3.2 Portability

Potable or drinking water is that which has its bacteriological, chemical and physical properties at acceptable levels thereby making it safe for drinking, cooking and other domestic purposes. Countries differ with their actual guidelines for quality water. This disparity contributes to the major challenges that developing countries has in ensuring potable drinking water [27]. The Guidelines seeks measure to support the development and risk implementation strategies that may ensure the provision of potable drinking-water through the management of unsafe constituents of water. These strategies may embrace national or regional standards developed from the scientific basis provided within the pointers. The guidelines are rules that describe minimum requisite of safe practice to guard the health of water end-users in order to derive numerical "guideline values" as pointers of constituents of water or indicators of water quality.

In order to outline obligatory limits, it's desirable to appraise the guidelines within the milieu of native or national, environmental, socio-economic and racial or ethnic conditions. The foremost reason for not advocating the adoption of international standards for drinking-water quality is the primacy provided by the inclusion of a risk–benefit approach (qualitative or quantitative) within the institution of national standards and laws. Further, the Guidelines for safe water practices are best carried out through associate degree integrated pre-emptive management substructure for safety applied from abstraction to end-usage. The guidelines offer a scientific locus of departure for national authorities to develop drinking water laws and standards applicable for the national state of affairs. In developing standards and laws, care ought to be taken to make sure that meager resources are not unnecessarily misappropriated (as the case with many developing countries) to other unproductive projects of national need quite unrelated or comparatively of minor importance to public health [28].

The approach in accordance with these guidelines is meant to steer to national standards and laws that may be promptly engaged and enforced to safeguard public health. The framework and methods of drinking-water guidelines could vary among countries and regions creating substantive gaps between developed countries and developing countries. This is because there's no single approach that's universally applicable. It is crucial within the development and execution of standards that the present and planned legislation with respect to water, health and native or local government are taken into consideration and that the tenacity to develop and enforce regulatory laws is assessed. Considering the unique political gaps and technological advancement within third world countries, different and workable approaches need to be test run and implemented when proven reliable. This is because standards that may work in one country or region may or may not be applicable in another country or region(s) owing to the fact that the assessment of safety of what the acceptable limits or risks are is peculiar to circumstances and is also a function of which society in part or as a whole has a functional role to play. It is therefore imperative that every country review its needs and abilities in developing a characteristic framework for water quality. Drinking water guidelines specifies maximum acceptable values/limits for a number of contaminants in drinking water. These values/limits are presented for minimum concentrations of microbial contaminants such bacteria, viruses, and parasites; chemical contaminants of health importance including both specific inorganic and organic constituents, pesticides, disinfectants, and disinfection byproducts; radioactive elements and substances and parameters in drinking water that may give rise to complaints from end users The ultimate judgment on whether or not the profit ensuing from the adoption of any of the rules and guideline values as national or local standards accounts for the cost is for every country to determine [27, 28].

Making water safe and potable is a big challenge in developing countries. To make water potable, it is usually put through to one or more treatment processes targeted at removing impurities in order to improve its portability and/or its pleasing quality. Various treatment processes exist for fresh water treatment. These include; coagulation, sedimentation, granular media filtration, adsorption, ion exchange, membrane filtration, slow sand filtration, and disinfection, and sometimes softening. Moreover, how these techniques and processes are applied in order to make drinking water more potable and safer in most developing nations is a major challenge because of lack of sufficient funds and other compounding economic issues. Potable drinking water must be among the highest priorities for every nation on earth because in recent times contaminated water kills more people than cancer, AIDS, wars or accidents and is mostly in developing nations. It is imperative that the water which humans drink be free of pathogenic organisms and toxic chemicals that constitute possible risk to public health. Moreover, developing countries constitute more than 70% of the world's human population technologies for making drinking water safe must be a pre-requisite in order to it accessible, affordable and environmentally sound, and as well geared towards becoming the nation's cultural norms [29].

Technological choices for fresh water treatment fall into two main categories – those that are used by urban authorities at clustered points from where water is then distributed, and those that are usually applied by individuals at homes even in rural areas. The most common and effective techniques for fresh water purification is chlorination. This is because this point-of- use (POU) water treatment strategy is cheap and easy to practice. But this is seldom rampant in most developing countries. There is a need for strong advocacy and serious engagement of this simple technology in the developing countries in order to curb the rising challenge of safe drinking water. Even under poor sanitary and hygienic conditions, in which people collect whatever water that is available from community tanks, wells, pumps and taps for use in their homes, if water is chlorinated, a dramatic decline in the incidence of water-borne diseases follows. Chlorination as a method has been in use for the treatment of public water supplies since the beginning of the twentieth century. The use of sodium hypochloride solution is a common method of chlorination that is used for the treatment of water in developing countries. The solution is placed in a capped bottle with directions for use. The user adds one full bottle cap volume of the solution to clear water (or two cups volumes for turbid water) to a standard-sized storage container. The user shakes the container and then waits 30 minutes before drinking. One major challenge of chlorination is the possible formation of carcinogenic disinfectant-by-products due to the presence of high organic components. But it's overwhelming popularity comes from the fact that it leaves a residual in the water matrix which helps in controlling possible re-growth of pathogenic microbes [30], although bacteria in drinking water distribution systems have devised a mechanism of evading chlorine disinfection through the formation of biofilm and acquisition of antimicrobial resistance elements [31].

In the island-cluster nation of Maldives, in the Arabian Sea/Indian Ocean, a technique has been developed that is cost effective and practicable at home by individuals for producing safe drinking water. This method is called solar water disinfection technique (or SODIS) with more than a dozen solar water disinfection techniques been perfected and put into practice. A process where sunlight and plastic containers are used to kill disease-causing organisms that contaminates water meant for domestic purposes. An individual or family, village or community may choose the technique that is more suitable to its circumstances [29].

The role of solar disinfection of water has many prospects in developing countries because of economic issues in these countries. Solar disinfection (SODIS) method is the commonly used method because of its cost effectiveness for disinfection of water meant for oral rehydration solutions [32]. In this method, about 0.3–2.0 liters of plastic soda bottles is filled with low turbid water. It is then shaken to oxygenate the water. After shaking, the bottles are allowed to stand under the sun for 6 hours and 2 days for a cloudy weather [33]. Organisms that are sensitive to this method include bacteria and viruses, protozoans such as *Cryptosporidium* and *Giardia* [34]. The innovations of using direct sunlight to disinfect water has been widely applied and the results has always been positive [35, 36]. Several challenges affects the use of this method because despite its usefulness and advantages (cheap technology), the major limitation in using this system still remains the maximum trapping of the sunlight for such purposes as the designing of solar systems are not always easily affordable in developing countries. Another challenge is the seasonal and weather changing factors that affect the intensity of the ultraviolet light coupled with the length of time it takes to treat only small volumes of water. In high turbid waters, a pre treatment using flocculation, sedimentation or filtration is usually employed before solar disinfection in order to obtain a more accurate result [31].

In general, several factors compound the easy practice of water treatment in the developing countries. These challenges to effective drinking water supply in developing countries include the natural scarcity of water in certain areas, floods, climate change, stratification and aeration of abstraction points, poor access and poor resource management.

Flood gives rise to source-receptor issues by increasing the concentration of silt thereby causing more siltation problems in river systems as well as the contamination of rivers and large dams. Climate change and water scarcity are also some of the concerns [37, 38]. Breaking down the thermoclines during stratification is also a challenge in that it requires much energy during abstraction and aeration of the source. Poor water productivity in the agricultural sector can impact on water quality [39]. Easy affordability, investment into water infrastructure and quality storage facilities to prevent contamination is the issues that need to be addressed. This entail qualitative education and awareness campaign after serious cross-contamination of its prospect [40–42]. The most important aspect in the maintenance of clean drinking water in the developing countries is an integrated approach of various feasible ways for achieving adequate water that is safe for public use. Proper management of solid waste and waste water can enhance the quality of drinking water systems [43, 44].

3.3 Availability & scarcity

Another prevailing challenge to safe drinking water in developing countries is the issue of natural scarcity of fresh water sources in certain areas especially in sub-Saharan Africa. For instance, at different times of the year, the fresh water availability varies. For example rivers are predominantly used during the wet season while underground such as borehole or dug-well water sources are used during the dry season [31].

Safe drinking water is a not just a privilege but a birthright of every individual irrespective of social class because it is connected to every facet of life on earth. It is also a major criterion for a robust national growth. Thus to maintain a robust and healthy society, adequate, reliable, clean, accessible, acceptable and potable drinking water supply has to be available for everybody.

Two factors dominate the scarcity of safe drinking water in developing countries. These are physical and economic scarcities. In some countries, *physical scarcity* where sufficient freshwater is not available is a leading challenge while in some countries (e.g. Nigeria, Ethiopia, Congo and Papua New Guinea, Ethiopia) where economic scarcity is the key challenge, abundant freshwater is available, but it is expensive to use. The other challenge is the fast increase in world's population and shifting demography at an alarming rate, whereas the available freshwater resources seldom change but almost remain constant [9].

Although environmental factors contribute to the issue of water scarcity, it is compounding because it may be mostly anthropogenic (man-made). More than two-thirds of the earth's surface is covered with water, but mostly exist as sea water and undrinkable unless desalinated. Only about 2.7% of this water is available as freshwater in lakes, brooks, ponds, stream, springs, groundwater and rivers on earth with only accessible proportion of 1%. This is because most of the available freshwater resources are inaccessible existing deep aquifers or frozen in the polar ice regions accounting for the very small proportion of safe drinkable water on earth (~3%) in the freshwater resources.

The issue of water scarcity in the third world is an ugly display of poor governance. Africa as continent should be the least with safe water challenge across the globe. With exemption of some water stressed countries such as South Africa, Morocco, Libya, Egypt, Niger, Tunisia, Algeria, Djibouti, Namibia, Sudan etc., other African countries with excess freshwater resources (e.g. Nigeria, Ethiopia, Congo and Papua New Guinea) still suffer from unavailability of adequate potable or safe drinking water due to economic reasons. South Africa receives about 450 mm annual rainfall and is classified as a water-stressed country with available freshwater resource that can only sustain 80 million people [45, 46]. On the contrary, Ethiopia being the second populous countries in Africa with more than eight river basins serving as the water tower of the whole east Africa due to the availability of surplus freshwater still struggles with count as the country among nations affected by long-standing water problem. The major cause of water scarcity in the world is further exasperated either by reduced water volume per area or the increasing demand for water due to population growth and the repudiating water quality by pollution [9].

There are two key factors over which the global concerns for the future availability of freshwater (safe drinking water in particular) centers on. They are the dramatic increase in the total withdrawals of freshwater in recent times and the incessant rise in the world's population across the globe especially in the developing countries. Currently, groundwater aquifers are undergoing repeated depletion even faster than they are been replenished in most parts of China, India and United States. This situation is mainly caused by unbalanced irrigation practices that not only plays significant role in groundwater exhaustion but in the degradation of soil quality thus reducing agricultural productivity making these region to be prone to future food insecurity and placing the future goal of 'green revolution' at jeopardy. Over the past 4-5 decades, freshwater withdrawals have doubled across the globe.

In the 3–10 years, the world population is estimated to reach 8–9 billion with a 1.5 billion projected yearly increase within the next 25–30 years. This will automatically reduce the availability of freshwater to 40% with respect to access to freshwater per person per year. As of date, reasonable number of the world's population (6 billion people) relies on freshwater supply and are currently facing water shortages. For instance in 31 countries (most developing countries such as Botswana, China, Ethiopia, Haiti, India, Kenya, Nigeria, Peru and Yemen) they are presently about 2.8 billion people that are confronted with the challenge of severe water problems.

According to the report by World Resources Institute [47], about 17 countries namely; Qatar, Israel, Lebanon, Iran, Jordan, Libya, Kuwait, Saudi Arabia, Eritrea, UAE, San Marino, Bahrain, India, Pakistan, Turkmenistan, Oman, and Botswana as of 2019 comprising about 1.7 billion people (about a fourth part of the world's population) are confronted with "Extremely High" water stress challenges. This challenge predisposes more than 1 billion people to water scarcity crisis and may reach about 3.5 billion people before 2025. Twelve among these countries majorly located within the MENA (Middle East and North Africa) region were ranked based on their exposure to water stress drought, flood and drought risk which are basic factors that controls water availability. India with the leading population (1.4 billion) among these 17 countries is expected to take the lead the effort in forming alliance with the rest of the 16 counterpart countries to demystify this impending global crisis because a country with such huge population is most vulnerable to freshwater shortages. If these issues are not tackled headlong, these countries could be at jeopardy of huge economic crisis resulting from climate related freshwater scarcity. They may end up losing about 6–14% GDP by 2050.

The challenge of safe water scarcity has raised some level of panic across the globe both at the local and international level, underpinned by the United Nations Sustainable Development Goals 6 (SDG-6) geared at providing access to clean drinking water for all and highlighting its importance to hygiene and the prevention of diseases. Despite these, inequalities remain in accessing safe drinking water in the world [48] (**Figure 1**).

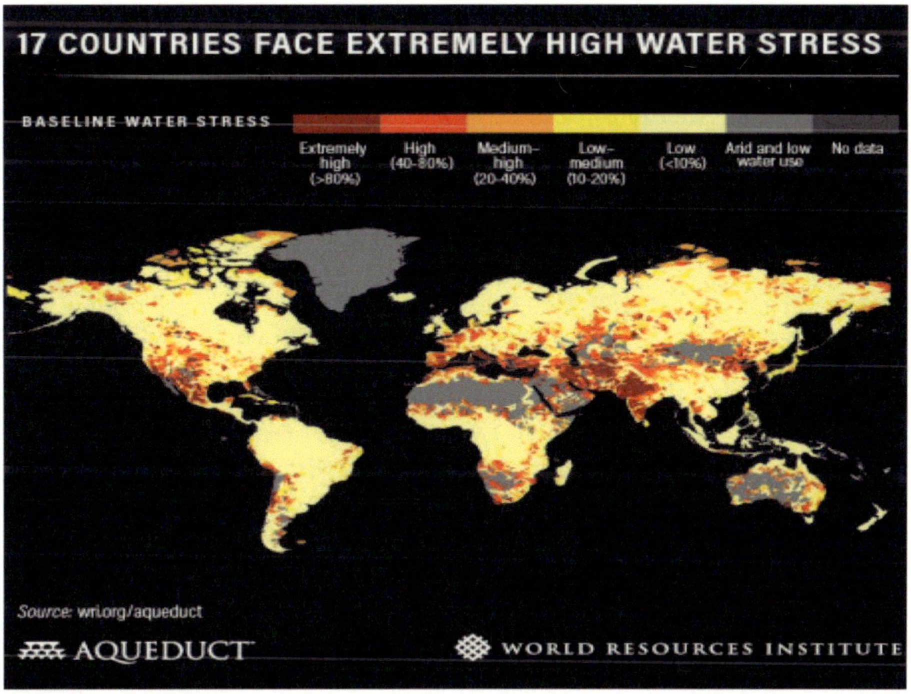

Figure 1.
Countries with extremely high water stress [47].

This challenge is worrisome because it has been observed that within a generation, the world's population may likely increase to about 8 billion people. But as we know, there is little or no significant rise in the amount of available freshwater as this population surges but will likely remain the same. It is as clear and enthralling as intact water thumbing down a mountain stream in a chain reaction. A novel approach must be taken in order to find new and objective ways of saving the situation by using and recycling the available freshwater at our disposal [29].

The problem of potable water shortage has become a source of concern in the world; locally and globally. Despite the concerted effort by the United Nations Sustainable Development Goals 6 (SDG-6) channeled at providing access to clean drinking water for all and underlining its import to hygiene and the prevention of diseases. Notwithstanding these efforts, inequalities remain in accessing safe drinking water in the world as more than 80 countries still suffer from water scarcity as it reflects on their public health and economic well being. In Nigeria, the origin of water shortage is dated as far back as the pre-colonial time when people had to journey from one part of the country to another in search of water.

The mismatch between demand for freshwater and its availability is a major causative factor of water scarcity although it may still arise from differences in human relationship (a social construct that culminates from product of affluence, presupposition and accordance with accepted societal standards, rules or expectations) or the resultant effect of modified supply patterns, rising from climate change. Water scarcity mostly affect people in the ghetto of developing countries as they are reported to, often, pay 5–10 times more per unit of water than people, in the urban areas, with access to pipe-borne water. This is because water scarcity is a relative concept and can occur at any level of supply or demand [49].

The availability of freshwater controls the bulk portion of the world economy. Technological advancement and industrialization are two other strong elements that threaten freshwater resources all over the world. In order to create balance, adequate provision of water is essential for agricultural purposes, drinking and cooking, industrial productions as well as recreation. Unfortunately, natural or artificial contaminants deny us of the right thus exposing us to a lot more challenging world. It is a well known fact that fresh water is an important necessity for our health [35, 50].

3.4 Accessibility & affordability

It is estimated safe drinking water is accessible by 89% of world population, while 11% do not have access to water suitable for drinking and cooking. Report from the "United Nation Development Programme", states that one out of six people do not have access to clean water, and this constitute about 1.1 billion people with lack or poor access to safe drinking water. On a normal basis, access to safe water in satisfactory amounts for drinking and domestic uses coupled with sanitation equipments that do not trade-off health or dignity is a need for every individual. The United Nations and other developed countries states that access to potable drinking water is an elementary human right, and it is a necessary step towards upgrading living standards globally. Unequal in access to potable water and hygiene are not morally acceptable, and are forbidden under international law. In some developing countries nearly 50% of the population do not have access to safe drinking water and hence, is characterize with poor health standard, especially in Africa, Middle East, Latin America and Asia. It is interesting to know that the number of people without access to safe drinking water is more than the number reported by the "United Nation Development

Programme". This is because there is little or no commitment in ensuring that most of the water supply facilities initiated during the MDGs in developing countries are functioning properly [9].

Equal access to safe drinking water according to the United Nation (UN) and other countries is a fundamental human right, and a key step towards enhancing living standards. Access to potable remains one of the main goal of Unite Nations-Millennium Development Goals (UN-MDGs) and also a major goal of the Sustainable Development Goals (SDGs). The importance of equal access to safe drinking water is further driven by goal 6 of The UN-SDG which states that "*Water sustains life, but safe clean drinking water defines civilization*". But in spite of these pertinent claims, unequal access to safe drinking water is still a major challenge in the world particularly in developing countries. The only way to curtail this problem is to tentatively address the issue of poor access to water and poor water resource management since equal access to potable drinking water is also considered to be a human right, not a privilege, for every man, woman and child. Inadequate water productivity in the agricultural sector can negatively affect water quality [31].

Unequal access to potable water has been a rising issue that confronts a large portion of the world's population both in developed and developing nations. Several factors could be responsible for that ranging from proximity to available freshwater pool, political, cultural or economic factor. This plays out in the variation of access to safe water both among and within countries. In Congo 77% of the population dwelling in towns and cities can access potable drinking water but on the contrary, only 17% of rural dwellers have access to safe drinking water. But in the Lao Peoples' Democratic Republic, the reversed is the case because almost every people in the rural area of Lao has access to potable drinking water compared to the only 60% of the overall residents that dwell in the capital city of Vientiane. Most importantly is the variation in the use of freshwater resources from one country to another.

In developing countries, almost 90% of freshwater is solely used for agricultural practices, 8% for industrial actions, while only 5% of their available freshwater are used for domestic purposes. But in developed countries, industrial use of freshwater is about 59%, 30% for agriculture and just 11% for domestic uses [29]. From the public health viewpoint, the amount of the population with equal access to safe drinking-water is the most vital single indicator of the overall success of a drinking-water supply scheme. There are various definitions to access to freshwater or supply coverage, with many of the definitions placing emphasis on safety or adequacy. The most accepted definition is that used by "reasonable access" to improved sources as being "availability of a minimum of 5.28 gallons per person per day within one kilometer of the user's dwelling from an improved source" [28].

The affordability of water has an important influence on the utilization of water and choice of water sources. Households with rock bottom levels of access to potable water system often pay a lot for their water than do households connected to a piped water system. The high price of water may push these households with least access to potable water to use to use different sources of water of water of poorer quality that represent a higher risk to health. Furthermore, high prices of water might scale back the volumes of water utilized by households that successively might influence sanitation and increase risks of disease transmission. When appraising affordability of potable water, it is pertinent to assemble data on the price at the point of purchase. However, if households are connected to the drinking-water supplier, this may be the tariff applied. But where water is bought from public standpipes or from neighbors, the cost at the point of purchase may be totally different from the supplier tariff.

Many different potable water sources (notably vendors) also involve costs, and these costs should be part of appraisals for affordability. In addition to periodic costs, the costs for elementary purchase of a connection should also be put in place when assessing affordability [28].

3.5 Sustainability

Developing countries face a huge challenge with sustainable safe water supply. While facing the challenges of the "old" agenda which are providing safe water services to every household encompassing coverage changed, rising cost of raw water, inefficient water utilities, poor public spending, inefficient private financing in the informal sector, excessive cost and economic impact on end users, lack of innovative and equitable, approaches to financing of water and sanitation services, equal access to both the poor and the rich, development of efficient formal institutions and adequate private sector investment in the water sector, developing countries are now facing the challenges of the new agenda.

The most eminent challenge faced by low-income and middle income countries are that of new agenda - developing quality ambient aquatic environment. While the quality of the aquatic environment is a primary concern in every country, the condition among cities in developing countries is exceptionally critical. In some middle-income countries wastewater is seldom treated. For instance in most Middle-income countries of Latin America, only about 2% of wastewater are treated. Data from the United Nations Environment Program revealed that water quality is extremely poor in developing countries compared to industrialized countries. Moreover, while the quality of aquatic environment in high-income countries improved over the 1980s; middle-income countries did not experience any improvement, whereas low-income countries experienced a sharp decline [51]. Other major challenges facing sustainable safe water supply include lack of sufficient capital investment and good governance, lack of appropriate policies and programs that consider rural diversity, infrequent and insufficient application of adaptive capacity indicators in urban sustainable water supply, unavailability of sustainable and resilient smart water grids in urban areas, lack of adequate water demand and supply modeling.

3.6 Climate change

Most developing countries are located in the world regions that are exposed the severe droughts and seasonal changes in precipitation and evaporation and this create instability and challenges the source of the water at different times of the year. Climate change and water scarcity are increasing challenge posing serious global concern because of they create a corkscrew debt burden on developing countries. Climate change affects the rate of extreme weather events and thus increasing the unpredictability about freshwater availability and reliability. The rise in global mean temperature will cause a widespread, high-speed, and intensifying climate change which will in turn play a significant role in changing the quality of drinking water (see **Figure 2**). In order to curb this impending catastrophe, professionals need to explore and adopt the method of better and efficient interrelated mechanisms between climate change and drinking water quality. This can also be achieved by adapting and optimizing the already existing water management practices in ensuring drinking water safety.

More emphasis should be given to pressing climatic issues such as adequate water governance for cross-border basins, trans-boundary information systems, a well

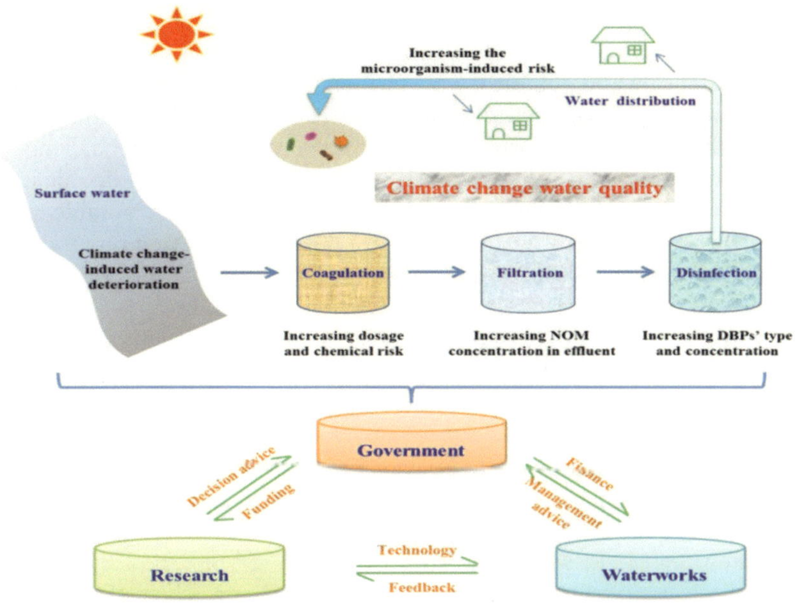

Figure 2.
Effect of climate change on drinking water safety [52].

established knowledge base schemes for mountain and valley terrains and sharing benefit between upstream and downstream communities. Floods has the capacity to create more siltation-like problems (eutrophication, sedimentation, land-sliding, acidification and salinization) in river systems as well as the pollution of rivers and large dams which will cause source receptor problems. Stratification issues in lake abstraction areas and ventilation of abstraction point to break down the water-table layers are needed but these processes requires much energy and resources [9, 31, 52].

Developing countries are vulnerable to end-to-end of normal climatic variations. It is advisable to pre-empty the impact of rise in the global mean temperature, accumulation of green house gases in the atmosphere and continuous depletion of the ozone layers because climate change it will probably increase the frequency and magnitude of some extreme weather conditions and disasters. Whether the world will quickly adapt to climate change is subject to current adaptive measures and the development of workable models that are currently been pursued by developing countries. Many frameworks are presently available for vulnerability and adaptation evaluation with both merits and demerits. But one major loophole in the approach adopted by the developing countries is that of investing resource that focuses on recovery from a disaster than on the creation of future adaptive capacity. There is need to urgently increase the capacity to manage harsh weather condition in order to reduce their weight on the economy, social and human damage and capital investments, thus shielding developing countries from frequent borrowing of money from the international lending agencies. Developing countries must also incorporate disaster management and adaptation as part of long-term sustainable development goals in combating extreme climate variation and water scarcity. On the other hands, international monetary agencies and financial donors must overhaul and repackage their agenda and investment policies in developing countries to basically focus on capacity development rather than tailoring their investment in recovery operations and infrastructural development [37].

4. Conclusion

Making water safe and potable is a big challenge in developing countries. More than 5 million people die yearly from water associated disease. Developing countries simultaneously face enormous financial, technical and institutional challenges in managing the quantity and quality of their water resources in a sustainable way. The high price of water may push these households with least access to potable water to use to use different sources of water of water of poorer quality that represent a higher risk to health. From the public health viewpoint, the amount of the population with equal access to safe drinking-water is the most vital single indicator of the overall success of a drinking-water supply scheme. Inadequate water productivity in the agricultural sector can negatively affect water quality. The problem of potable water shortage has become a source of concern in the world; locally and globally. Thus to maintain a robust and healthy society, adequate, reliable, clean, accessible, acceptable and potable drinking water supply has to be available for everybody. By giving priority to an enhancing and well planned, structured and governed and regularly supervised infrastructure coupled with related institutional capacities the challenge of seasonal climatic alterations can be well curtailed.

Acknowledgements

We are grateful to God for making this project a success. Our heartfelt appreciation goes to my research team at Department of Microbiology, Evangel University Akaeze, Ebonyi State, Nigeria or their support and encouragement during the time of this work. We also acknowledge the team leader of this project Tea Jurcic and the entire team at IntechOpen for their support and encouragement throughout this work. Thank you all.

Author details

Collins Onyebuchi Okafor[1]*, Ude Ibiam Ude[2], Felicia Ngozi Okoh[1] and Blessing Osose Eromonsele[1]

1 Department of Microbiology, Evangel University Akaeze, Ebonyi State, Nigeria

2 Department of Applied Microbiology, Ebonyi State University, Abakaliki, Nigeria

*Address all correspondence to: cofor1@icloud.com; microprof@evangeluniversity.edu.ng

References

[1] Schwartz HE, Emel J, Dickens WJ, Rogers P, Thompson J. Water quality and flows. In: The Earth as Transformed by Human Action. Cambridge, UK: Cambridge University Press; 1990. pp. 253-270

[2] Brenniman GR. Potable water. In: Environmental Geology. Encyclopedia of Earth Science. Dordrecht: Springer; 1999. DOI: 10.1007/1-4020-4494-1_265

[3] Okafor COO, Ifeanyichukwu RI, Ibiam UU, Stanley CO, Chika E, Kenneth NO, et al. Drug resistance profile of biofilm forming Pseudomonas aeruginosa isolated from aquatic environment in South Eastern Nigeria. Environmental Challenges. 2022;**8**:100530. DOI: 10.1016/j.envc.2022.100530

[4] Falkenmark M. Growing water scarcity in agriculture: Future challenge to global water security. Philosophical Transactions of the Royal Society A. 30 Sep 2013;**371**(2002):20120410. DOI: 10.1098/rsta.2012.0410

[5] Ahile SI, Udoumoh EF, Adzande P. Residents coping strategies with water scarcity in Makurdi town, Nigeria. Mediterranean Journal of Social Sciences. 2015;**6**(4):S2

[6] Bos R, Alves D, Latorre C, Macleod N, Payen G, Roaf V, et al. Manual on the Human Rights to Safe Drinking Water and Sanitation for Practitioners. London, UK: IWA Publishing; 2016

[7] Falkenmark M, Rockstrom J. Balancing Water for Human and Nature: The New Approach in Ecology. London: Earthscan Publication; 2004. p. 320

[8] Mpenyana-Monyatsi L, Momba MNB. Assessment of groundwater quality in the rural areas of the North West Province, South Africa. Scientific Research and Essays. 2012;**8**(7):903-914

[9] Dinka MO. Safe drinking water: Concepts, benefits, principles and standards. In: Glavan M, editor. Water Challenges of an Urbanizing World. London: IntechOpen; 2018. DOI: 10.5772/intechopen.71352. Available from: https://www.intechopen.com/chapters/57345

[10] WHO. UN-Water Global Annual Assessment of Sanitation and Drinking Water Report: The Changes of Extending Sustaining Services, UN Water Report 2012. Switzerland: WHO; 2012

[11] OECD. Benefits of Investing in Water and Sanitation: An OECD Perspective. OECD Studies on Water. Paris: OECD Publishing; 2011. p. 148. DOI: 10.1787/9789264100817-en

[12] Biswas KA. Integrated water resources management: A reassessment a water forum contribution. International Water Resources Association. 2004;**29**(2):248-256

[13] Desbureaux S, Richard D, Aude-Sophie R, Jason R, Esha Z. The Impact of Water Quality on GDP Growth: Evidence from Around the World. Washington, DC: World Bank; 2019

[14] Alan H, Xavier L, Jack M, Diego R. Economic development. In: Josefina M, Carlos MG, Colin G, editors. Water for a Sustainable World. The United Nations World Water Development Report. United Nations Scientific and Cultural Organization (UNESCO). Paris, France: UNESCO; 2015

[15] WHO/UNICEF Joint Water Supply and Sanitation Monitoring Programme,

Millennium Development Goals. Drinking water: Equity, safety and sustainability. In: Drinking Water Equity, Safety and Sustainability: Thematic Report on Drinking Water 2011. Geneva/ New York: WHO/UNICEF; 2011. pp. 47-53

[16] Davis ML. Water and Wastewater Engineering: Design Principles and Practice. New York: McGraw-Hill Education; 2013

[17] Connor R, James W. The water–energy nexus. In: Michela M, Richard C, editors. Water and Energy. Vol. 1. United Nations World Water Assessment Programme (WWAP). The United Nations World Water Development Report. Paris: UNESCO; 2014. pp. 12-21

[18] de Araújo LCA, de Oliveira MBM. Effect of heavy metals on the biofilm formed by microorganisms from impacted aquatic environments. In: Dincer S, Özdenefe MS, Arkut A, editors. Bacterial Biofilms [Internet]. London: IntechOpen; 2019. DOI: 10.5772/intechopen.89545. Available from: https://www.intechopen.com/chapters/69536

[19] Critchley MM, Cromar NJ, McClure NC, Fallowfield HJ. The influence of the chemical composition of drinking water on cuprosolvency by biofilm bacteria. Journal of Applied Microbiology. 2003;**94**:501-507

[20] Environmental Protection Agency (EPA). Sulphate in Drinking Water. 2011 Available from: water.epa.gov/drink/contaminats/unregulated/Sulphatecfm#cdc [Accessed: January 10, 2019]

[21] Okafor COO, Nwadiogbu IA, Agumah NB, Eni OE, Odinkemere SC. Presence of multi drug resistant coliform bacteria isolated from biofilm of sachet and borehole waters sold in abakaliki metropolis, Ebonyi State, Nigeria. International Journal of Scientific and Technology Research. 2015a;**4**(6):59-64

[22] Okafor COO, Iroha IR, Nwadiogbu IA, Agumah NB, Eni O, Odinkemere SC. Antimicrobial resistance pattern of coliform bacteria isolated from sachet and borehole waters sold in Abakaliki metropolis of Ebonyi State, Nigeria. International Journal of Innovation Science and Research. 2015b;**16**(2):526-532

[23] Fawell J, Nieuwenhuijsen MJ. Contaminants in drinking water. British Medical Bulletin. 2003;**68**:199-208

[24] Agah VM, Oko C, Udu-Ibiam EO, Nwachi AC, Okonkwo EC, Okafor COO. Microbiological quality of natural spring waters in Enugu and Ebonyi state, south/East Nigeria. Journal of Biology, Chemistry and Pharmacy. 2018;**2**(1):1-7

[25] Sharma S, Bhattacharya A. Drinking water contamination and treatment techniques. Applied Water Science. 2017;7:1043-1067. DOI: 10.1007/s13201-016-0455-7

[26] Ghrefat H, Nazzal Y, Batayneh A, Zumlot T, Zaman H, Elawadi E, et al. Geochemical assessment of ground water contamination with special emphasizes on fluoride, a case study from Midyan Basin, north Western Saudi Arabia. Environment and Earth Science. 2014;**71**:1495-1505

[27] Gadgil A. Drinking water in developing countries. Annual Review of Energy and the Environment. 1998;**23**:253-286

[28] WHO. Guidelines for Drinking-Water Quality [Electronic Resource]: Incorporating 1st and 2nd Addenda. 3rd ed. Vol. 1. Recommendations; 2008

[29] Rao CNR, Prasad LV. Third world academy of sciences (TWAS). In: Daniel S, editor. Safe Drinking Water: The Need, the Problem, Solutions and an Action Plan. Trieste, Italy: TWAS; 2002. p. 23

[30] Arnold BF, Colford JM. Treating water with chlorine at point of use to improve water quality and reduce child diarrhea in developing countries: A systematic review and meta analysis. The American Society of Tropical Medicine and Hygiene. 2007;**76**(2):354-364

[31] Treacy J. Drinking Water Treatment and Challenges in Developing Countries. IntechOpen: Department of Applied Science, Limerick Institute of Technology, Limerick City, Ireland; 2019. DOI: 10.5772/intechopen.80780

[32] Acra A, Karahogopian Y, Raffoul Z. Solar Disinfection of Drinking Water and Oral Rehydration Solutions – Guidelines for Household Application in Developing Countries. Lebanon: UNICEF Beirut; 1984

[33] Mc Guigan KG, Conroy RM, Mosler HJ, DuPreez M, Ubomba-Jaswa E, Fernandez-Ibanez P. Solar water disinfection (SODIS); A review from bench-top to roof-top. Journal of Hazardous Materials. 2012;**235-236**:29-46

[34] Sommer B, Mariño A, Solarte YS, Salas ML, Dierolf C, Valiente C, et al. SODIS: An emerging water treatment process. Aqua (Oxford). 1997;**46**(3): 127-137

[35] Simons R, Gabbai UE, Moram MA. Optical fluence modelling for ultraviolet light emitting diode-based water treatment systems. Water Research. 2014;**66**:338-349

[36] Yumu LG, David RD, Corkish R, Ashbolt N, Jagals P, Stuetz R. Photovoltaic powered ultraviolet and visible light-emitting diodes for sustainable point-of-use disinfection of drinking waters. Science of the Total Environment. 2014;**493**:185-196

[37] Mirza MMQ. Climate change and extreme weather events: Can developing countries adapt? Climate Policy. 2003;**3**(3):233-248

[38] Vairavamoorthy K, Gorantiwar SD, Pathirana A. Managing urban water supplies in developing countries-climate change and water scarcity scenarios. Physics and Chemistry of the Earth parts A/B/C. 2008;**33**(5):330-339

[39] Cosgrove WJ, Loucks DP. Water management current and future challenges and research directions. In: Water Resource Research. Washington DC: American Geophysical Union (AGU) Publications; 2015

[40] Rodriguez D, Den Berg CV, McMahon A. Investing in water infrastructure: Capital, operations and management world bank water papers. 2012;**1**:1-52

[41] Ghosh S, Morella E. Africa Water and Sanitation Infrastructure Directions in Developments Infrastructure Access Affordability and Alternatives. Washington DC, USA: The World Bank; 2011

[42] Sobsey MD, Handzel T, Venczel L. Chlorination and safe storage of household drinking water in developing countries to reduce water borne diseases. Water Science and Technology. 2003;**47**(3):221-228

[43] Blueriver (2018). Water Quality and Water Resources Management. Available from: https://bluerivers.kiev.ua/water-quality-and-waterresources-management [Accessed: April 1, 2018]

[44] Thompson T, Sobsey M, Bartram J. Providing clean water, keeping water clean an integrated approach. International Journal of Environmental Health Research. 2003;**13**(1):S89-S94

[45] Department of Water Affairs (DWA). Integrated Water Resource Planning for South Africa: A Situation Analysis. Pretoria: Department of Water Affairs; 2011

[46] Binns T, Illgner P, Nel E. Water shortage, deforestation and development: South Africa's 'Working for Water' programme. Land Degradation and Development. 2001;**12**:341-355

[47] World Resource Institute (WRI). Countries with Extremely High Water Stress, 2019. Available from https://www.wri.org/insights/17-countries-home-one-quarter-worlds-population-face-extremely-high-water-stress [Accessed: September 9, 2022]

[48] Schyns JF, Hoekstra AY, Booij MJ. Review and classification of indicators of green water availability and scarcity. Hydrology and Earth System Sciences. 2015;**19**(11):4581-4608

[49] Iduseri EO, Abbas II, Izunobi JU, Ogedegbe SO, Ogbonna DO. Access to safe drinking water in developing countries: A comparative analysis of the urban and rural areas of Zaria, Kaduna State, Nigeria. Journal of Applied Sciences and Environmental Management. Manage. 2021;**25**(11):1869-1878. DOI: 10.4314/jasem.v25i11.4

[50] Resch M, Lavie É. Issues associated with drinking water supply in cities. EchoGéo. 2021;**57**. DOI: 10.4000/echogeo.22563. Available from: http://journals.openedition.org/echogeo/22563

[51] Briscoe J. The changing face of water infrastructure financing in developing countries. International Journal of Water Resources Development. 1999;**15**(3):301-308

[52] Baiwen M, Chengzhi H, Junya Z, Mathias U, Stefan P. Impact of climate change on drinking water safety. ACS EST Water. 2022;**2**:259-261

Chapter 10

Evaluation of a Landscape Irrigation Management Strategy to Support Abu Dhabi Update Its Water-Related Standards

Geraldine Seguela, John Richard Littlewood and George Karani

Abstract

This chapter discusses an landscape irrigation (LI) strategy to enable 100% non-potable water reuse through soil improvement, thereby reducing the environmental impacts. The case study site is a medical facility including 33,257 m^2 of landscaping in Abu Dhabi (AD), the capital of the United Arab Emirates. The aim of this research is to increase net-carbon sinks, a pillar of decarbonization, as the basis for a proposed protocol to implement soil improvement techniques for the landscape architecture/agriculture industries. The interventions, based on AD soil and water recycling standards, included three different soil additives in 2016 and 2017, together with the calculation and implementation of a suitable irrigation rate to establish LI demand and reduce a five-month shortfall in air-conditioning condensate water supply. The intervention results show the case study irrigation rate was 50% less after soil improvement than the AD Municipality irrigation standard and that the LI condensate water deficit decreased by 8046 m^3, a 42% reduction. The research demonstrates that carbon sinks can be increased through improved soil management; this highlights the need to update AD's water-related standards to help the city achieve its 2030 target of a 22% reduction in greenhouse gas emissions.

Keywords: water policy, climate change, water quality criteria, recreational water, bacteria, sodicity, salinity, soil acidity, irrigation and drainage, heavy metals and metalloids, soil interpretations

1. Introduction

1.1 Research background

The work documented in this chapter forms part of the first author's Professional Doctorate in Engineering research, a change project implemented at a medical facility case study (MFCS) in use since 2015. The MFCS is a 364-bedroom hospital located in Abu Dhabi (AD), the capital of the United Arab Emirates (UAE), a hot, desert-type climate as classified by Köppen and Geiger [1]. The 33,257 square meters

(m^2) landscape at the MFCS represents 50 percent (%) of the MCFS footprint and uses air-treated condensate water (CW), a product of air conditioning, for outdoor irrigation purposes to avoid the usage of energy-intensive desalinated water [2]. Due to peak CW formation occurring in summer, there is a shortfall in winter (established at −19,235 cubic meters per year (m^3/year) in 2016). Water and soil data were used to develop sustainable water consumption and reuse (SWC) strategy forming the basis of a water conservation protocol [3] whereby soil improvement for the landscape is investigated as part of a mixed methods approach [4–6]. This strategy would enable the MFCS to address the five-month CW shortfall, reduce the outdoor use of desalinated water, and, consequently, would reduce the MFCS's building systems water and energy consumption, operation and maintenance cost and practices, and ultimately greenhouse gas (GHG) emissions. The outcome of the research project demonstrated that carbon sinks can be increased through improved soil management [7], helping AD achieve its 2030 GHG emissions target [8].

1.2 The context of water: Soil Nexus in Abu Dhabi

Of the world's 19 most water-scarce countries, 13 are Arab countries [9, 10]. Per capita water availability is below 200 m^3 per year in eight Middle Eastern countries (**Figure 1**), including the UAE [12]. In December 2020, the UAE ranked 10th on the list of the 17 most water-stressed countries in the Middle East and North Africa (MENA) region [12].

AD's annual water consumption was estimated to be 2.49 billion m^3 in 2017 [13], and peak demand is predicted to more than double by 2030 [14]. In response, AD has embarked on a USD 5 billion programs based on an aquifer storage and recovery approach [15]. This program aims to build capacity so that local aquifers can be used as strategic reserves for desalinated water [16].

Analysis of water demand by sector and by type of water in AD shows that 100% of potable water is used for commercial, residential, and industry buildings, including

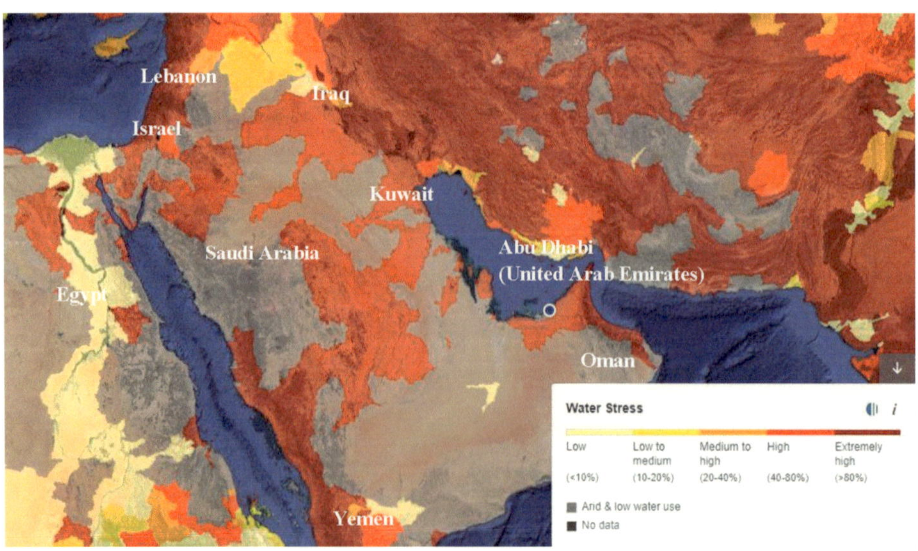

Figure 1.
Water Stress in the Middle East including the UAE by 2030 [11].

outdoor landscape irrigation (LI) [17]. Thus, a significant opportunity exists to conserve water for outdoor use.

Exacerbating the issue of water availability, the landscape of AD is dominated by sandy, salty soil with very low water-holding capacity and experiences high temperatures and relative humidity, limiting plant growth [18–20] and locally grown food [21]. In arid and semi-arid regions, soil qualities are frequently physically, hydraulically, and chemically deficient due to their sandy natural state and exposure to harsh climatic conditions [14, 19, 22].

Consumptive water use is the sum of two factors: transpiration and evaporation [23]. While agriculture is responsible for the bulk of water consumption—through both evaporation from land surfaces during irrigation and transpiration from plants [24]—conservation techniques for urban landscape are characterized by site modifications, diverse use, and complex microclimates [25]. In AD, the greatest potential for GHG emission reductions by 2030 is from combined electricity and water production (22% of business as usual (BAU) emissions), through tariff reform, building and efficiency standards, demand-side management, and district cooling and appliance efficiency standards [8].

1.3 Case study background

In 2016 and 2017, the MFCS Energy Monitoring and Control System (EMCS) recorded LI water deficits (predominantly occurring in the winter and spring months) of 19,235 and 11,189 m^3, respectively, representing a 42% year-on-year reduction. The use of CW and desalinated makeup water for LI decreased by 8% from 37%, from 91,564 to 83,960 m^3, over the same period. Additionally, total LI consumption of the combined air handling unit air conditioning (AHU A/C) CW and desalinated makeup water decreased by 18% from 2016 to 2017. These results derived from the implementation of a series of interventions through an action research methodology including soil enhancement, soil quality testing against the Regulation and Supervision Bureau standard [26] and the Ministry of Climate Change and Environment soil standards [27], as well as valve flow audit and water demand calculations based on Abu Dhabi Municipality [28], Urban Planning Council [29, 30], and the United States Environment Protection Agency standards (U.S. EPA) [31, 32]. A pilot empirical project (PES 2016) was undertaken to verify that the application of a soil conditioner could improve (i) plant growth, and (ii) water retention in the soil, thereby assisting in reducing water demand by up to 50% against the U.S. EPA standard [32], delivered by a 5% adjustment to the soil.

1.4 Gap analysis

In relation to the knowledge gaps in practice—identified and summarized in **Table 1** below, coupled with the pilot empirical study (PES 2016) findings—it was found that the AD soil and water standards conflict with or disregard each other or lack clear directions for water savings in landscape irrigation (LI). For instance, the way the Regulation and Supervision Bureau (RSB) [26] regulates LI is by including criteria for trace elements; however, no concentration values are defined for salinity or essential nutrients. This could be beneficial to irrigation water, plants, and soils [34, 35, 39]. It was also found that the Ministry of Climate Change and Environment (MOCCAE) [27] soil standard does not clearly indicate minimum and maximum soil micro- and macronutrient concentration limits for soil maintenance. In addition,

Existing AD codes, standards, and strategies	Local water-related conservation regulations gaps
Guide to Recycled Water and Biosolids Regulations 2010 [26]	Soil quality standard for water conservation: Beyond the potential for hydrogen (pH) and copper, RSB does not offer other parameter limits such as electrical conductivity, sodium, calcium, and magnesium concentrations, which influence water-holding capacity in soil [33].
Ministerial Resolution 476 of 2007 concerning by-law of fertilizers and agricultural soil conditioners, Chapter two [27]	Soil quality standard for water conservation: Ministry of Climate Change and Environment (MOCCAE), formerly known as Ministry of Environment and Water (MoEW), does not offer parameters limits such as electrical conductivity, sodium, calcium, and magnesium concentrations, which influence water holding capacity in soil [34, 35].
Design Public Realm Guideline [29]	This document promotes irrigation interval days without mandating soil amendment with soil additives.
Irrigation Systems and Operation Maintenance volume 2C Section 02850 [28]	This document addresses irrigation rates for design and construction projects, but not for building operation. And the irrigation recommended is not in line with vegetation watering recommendations [36, 37].
International Building Code 2013, Chapter 29 Plumbing Systems [38]	Building hydraulics, such as landscape irrigation valve flow audit, is not addressed.

Table 1.
Local standards and policies gap analysis of water conservation in Abu Dhabi.

the Urban Planning Council (UPC) [29] guideline conflicts with Abu Dhabi Municipality (ADM) standard [28] with regard to seasonal months and landscape irrigation rates. Last, the only common ground of Department of Municipal Affairs and Transport (DMAT) [38] and UPC Public Realm Design Manual [29] guidelines for water conservation in landscaping is the regulation of reusing treated sewage effluent (TSE) and *Legionella*. None of the above standards address a common and clear water demand management strategy for landscape irrigation, a pillar of decarbonization [7].

1.5 Contributions to new body of knowledge in practice

The key original and significant contributions to the knowledge gap in practice from this research chapter are threefold.

Firstly, improving the management of water and soil increases net-carbon sinks [40] and offsets GHG sources (e.g., fossil fuel), because some CO2 will be returned from the natural carbon sinks [7], such as ocean and soil acting as CO_2 absorbers. Secondly, the soil in AD is classified by the U.S. Department of Agriculture [18] as having a very low water-holding capacity. Water-holding capacity and infiltration rates are ultimately linked [34]. Infiltration is the most important factor in the soil phase of the hydrological cycle [41]. The infiltration rate is measured according to the soil's ability to absorb irrigation water [42]. Thirdly, the choice of essential plant nutrients to include in applied irrigation requires considerable professional skill and experience, as hydrology and pedology are complex sciences [23, 34, 35]. The problem related to water need and environmental conditions arises when an essential plant nutrient element is needed, and water is not needed [27]. Irrigation is a water consumptive process, which is also energy-intensive to deliver [25]. Hence, in identifying ways to help minimize water wastage and decrease GHG emissions for a MFCS in AD, this chapter addresses gaps in practice. Recommendations for revision of the AD water and soil standards are introduced in Section 1.4 above and described below.

2. Materials and methods

The MFCS (Water Resources case study) uses a quantitative data collection process [4–6] which links to one of the overall research objectives of the research project: to test a water conservation framework for water resources through two interventions and two calculations, as described below in **Table 2**, and to analyze the data collected.

2.1 Pilot empirical study (PES) 2016

As per **Table 2**, a pilot empirical study (PES 2016) was conducted from April to September 2016, which comprised four elements:

i. Soil quality testing before and after soil solution applications A, B, and C including a hydro-activator fertilizer combined with compost and synthetic fertilizer (solution A), an organic soil conditioner (solution B), and an organic compost combined with a synthetic fertilizer (solution C). These three different soil solutions were applied to 8855 m^2 (out of 24,402 m^2) of the MFCS landscape area between April 2016 and September 2016 to evaluate the suitability of soil improvements for reducing the use of potable water for outdoor use and its effect on plant growth. The purpose of the comparison was to allow measurement of the relative effect of the soil conditioner treatment (independent variable) against the areas that received either no treatment (no water reduction) or a different treatment (Solution A and C) with a different water reduction.

SOIL enhancement interventions	WATER demand calculations
Pilot Empirical Study: PES 2016	**Pilot Calculation One: PPC1—2016**
• Soil enhancement trial: April–September 2016 including the following:	• Irrigation rate calculations pilot
o Soil quality testing (March 2016)	**Calculation One: Calc 1–2017**
o Outdoor Valve Flow Audit Trial (December 2016)	• Irrigation rate calculation implementation
o Weekly Photographs	
o Energy Management Control System (EMCS) LI water consumption records	
Soil Intervention: SEI 2017	
• Soil enhancement implementation (April–May 2017) including the following:	
o Soil Quality Testing (June 2017)	
o Outdoor Valve Flow Audit (August 2017)	
o Weekly Photographs	
o EMCS records LI water consumption records	

Table 2.
Case study methodology summary.

ii. Audit of eight water valves out of 127 (connected to the irrigation controller and the LI water meters) to verify their flow accuracy (in l/m^3).

iii. Weekly photographs of the areas tested before and after the test to provide evidence of plant growth.

iv. EMCS records to monitor water consumption.

The intent of the pilot project (PES 2016) was to create and measure a credible outcome based on a cause-effect relationship within the research field setting (MFCS PES 2016 landscape area) [33]. Findings from this intervention have been published in Seguela et al. [33].

2.2 Soil enhancement implementation (SEI) 2017

The SEI 2017 intervention was conducted to reduce the water demand strategy onsite, following the successful implementation of the PES 2016, Pilot Project Empirical Study, Solution B.

As a result of the PES 2016 analysis, soil solution B was applied to the remaining of the MFCS landscape area (33,257 m^2–8855 m^2 = 24,402 m^2) including date palms, plants, shrubs, groundcover, and lawn. As per PES 2016 application [33], soil conditioner B was made up of 55% organic matter, 10% biodegradable polymer, humid acid, 5% sulfur, and 1% nitrogen (Solution B manufacturer information).

A total of six soil specimens were randomly sampled on site in June 2017 as per PES 2016 soil sampling method (adapted from Britton Harrell [43] in Seguela et al. [33]), before applying Solution B to the remaining 24,402 m^2 of the MFCS landscape to establish soil quality improvement against 2016. The major soil macronutrients and major micronutrients, listed in **Tables 3** and **4**, were tested [23, 44, 45] to compare SEI 2017 Intervention soil quality results against PES March 2016 results [33] as per the method described in Section 2.2. The 2017 soil test results were evaluated against Hornek et al.'s [45], Jensen's [46], and Flynn and Ulery's [47] concentration limits.

For this chapter, the results will firstly be assessed for the 11,000 m^2 of *Pennisetum setaceum* only, which are classified as low-irrigation plants by UPC [29, 30] and require 1.5 kilograms per square meter (kg/m^2) of soil conditioner B. Secondly, the results will provide an account of the remaining 127 outdoor valves water flow audit (135–8 = 127 valves). Thirdly, the results will visually be evidenced.

2.3 Updated pilot calculation one method (PPC1 2016): irrigation rate

2.3.1 Updated literature: UPC (2017) irrigation rate standard

During the construction stage of the MFCS in 2011, the LI demand was estimated by the landscape contractor at 375 m^3/month at peak time (July) [33] based on ADM standard irrigation rate [28]. In 2017, the ADM Standard [28] was announced [36, 48] in Seguela et al. 2017 [33] to be adjusted and aligned to UPC [30] irrigation rate, which was updated and published in August 2017. From the analysis of ADM [28] and UPC [30] irrigation rates, it is observed that UPC [30] revised irrigation rate is 19% lower than the original [29].

The revised UPC manual [30] reflects the findings of Seguela et al. [33], which recommend that when applying both soil conditioner and organic fertilizer the

Soil samples	Year	OM (%)	MC (%)	Ca (mg/kg)	Mg (mg/kg)	K (mg/kg)	N (mg/kg)	Na (mg/kg)
Sample 1	2016	6.16	8.1	40.08	24.31	23.25	50	—
	2017	1.89	2.26	9.62	8.26	15.1	12.3	18.3
Sample 2	2016	1.88	10.7	40.08	29.17	48.7	60	—
	2017	1.77	3.12	7.2	5.3	7.04	2.94	8.98
Sample 3	2016	3.74	5.7	160.32	53.48	34.15	70	—
	2017	2.33	1.34	28	12	8.06	14.35	14.34
Sample 4	2016	2.96	3.5	53.44	32.4	7.35	20	—
	2017	2.17	2.93	44	28	21.16	20.06	38.94
Sample 5	2016	2.06	13	24.04	14.6	4	40	—
	2017	1.21	1.71	29	20	16.34	13.89	18.55
Sample 6	2016	—	—	—	—	—	—	—
	2017	1.89	2.15	26	12	7.07	15.06	12.44

Notes: Organic Matter (OM), Moisture Content (MC), Calcium (Ca), Magnesium (Mg), Potassium (K), Nitrogen (N), Sodium (Na).

Table 3.
SEI 2017 against PES 2016 soil laboratory test results.

Soil samples	Year	pH	EC (ds/m)	SAR (meq/l)	ESP (%)
Sample 1	2016	8.27	1.14	4.02	59
	2017	7.08	0.25	0.89	35.39
Sample 2	2016	8.72	1.00	5.96	63
	2017	7.04	0.18	0.62	31.56
Sample 3	2016	7.93	1.23	0.35	7.5
	2017	7.27	0.31	0.57	22
Sample 4	2016	8.05	0.33	0.66	21
	2017	7.34	0.74	1.13	29.12
Sample 5	2016	8.17	0.17	21.4	93
	2017	7.42	0.50	0.65	21.69
Sample 6	2016	—	—	—	—
	2017	7.46	0.31	0.51	21.06

Notes: Electrical conductivity (EC), sodium absorption ratio (SAR), exchangeable sodium percentage (ESP).

Table 4.
SEI 2017 against PES 2016 soil salinity laboratory test results.

water- and nutrient-holding capacity of the soil increases and water needs decrease while sustaining plant growth. Yet, UPC [30] advocates irrigation application every 3 days during the summer and every 7 to 10 days during winter, contrary to the results of Seguela et al. [33]. See results in Section 3.1 below. It is interesting to note that the UPC revised manual [30] removed the interval days (ID) recommendations within the irrigation rate schedules at pages 162 and 163 but has kept the ID within the text of the manual.

2.3.2 PPC1 2016 against Calc1 2017: Irrigation rates comparison

Pilot irrigation rate calculation one (PPC1 2016), calculated by the author [33] and adapted from U.S. EPA [31, 32], is based on decisive parameters affecting the irrigation requirements [33].

The irrigation budget method [31, 32] is based on the peak irrigation month (Landscape Water Requirement for July) and has been compared with the local irrigation rate from the ADM [28] and UPC [29, 30]. The UPC [29] and ADM [28] standards did not have the same seasonal patterns. For instance, the UPC defined winter as the period from January to March (3 months), whereas the ADM considered it to extend from December to March (4 months). The Pilot Calculation One (PPC1 2016) method follows ADM [28] seasonal patterns because the UPC's new manual [29] no longer includes seasons. The UPC [29] irrigation rate was based on seasonal water reduction by interval day's irrigation patterns according to season, plant maturity, and type of plants. For instance, in July (peak month) the UPC [29] recommended *P. setaceum* be irrigated at a rate of 10 liters per m^2 applied every 2.5 days. The updated UPC irrigation rate [30] recommends 10.2 liters per m^2 every 3 days.

The ADM [28] irrigation rate was based on seasonal water reduction according to seasons and type of plants, which, according to the landscape contractor, was 20 liters per m^2 per day for *P. setaceum* in summer. In comparison, Lee [36] recommends 12 liters per m^2 per day.

In 2018, the UPC [30] irrigation rate was still not mandatory for Operations and Maintenance (O&M) projects not having been rated with the Estidama [48] Pearl Building Rating System [33]. Since ADM is the primary authority for approving construction building permits, the MFCS landscape contractor is following the ADM [28] standard for maintenance projects [49] even though irrigation rate application at operations is not a mandate [28–30, 48, 50].

3. Results

3.1 Reflection on PES 2016 soil enhancement results

In 2016, it was observed the landscape irrigation controller installed at the MFCS was not compatible with the Energy and Management Control System (EMCS) and the two were disconnected systems. The irrigation controller's only function is to provide adequate water quantity to the plant at a defined time. The flow meters monitoring overall LI consumption are connected to the EMCS. It would have been preferable to install flow sensors in addition to the flow meters connected to the EMCS for each valve connected to the irrigation controller to monitor LI consumption. This also may have helped to detect any outdoor water leakage [51, 52], because the landscape contractor would have had direct access to the water reports. For financial reasons, this solution was not pursued by the MFCS. Thus, it was not possible to establish the exact quantity of water saved for the pilot project in relation to the application of solutions A, B, and C.

During the pilot project implementation, solution B water consumption reduction could not be evidenced due to the inaccurate irrigation controller settings (from the valve flow errors) and the inaccurate reading of the uncalibrated flow meters connected to the EMCS. This said, these findings have helped the project team to reevaluate the overall irrigation demand so that the soil enhancement pilot could be truly quantified based on the accurate valve flow and on the EMCS calibrated flow meters [33].

When using ultrapure water, such as air-conditioning condensate water, for landscape irrigation (LI), maintaining optimum soil nutrient levels (primary, secondary, and micronutrients) can be challenging, especially in the first few surface centimeters of the soil [34]. The tendency of the low-salinity water is to strip cations from exchange sites [34, 35, 46, 53]. The stripping process can affect both plant-available nutrients and cations (calcium, magnesium, sodium, potassium, and hydrogen) needed to preserve soil structure. Soil conditioner B increased the Cation Exchange Capacity (CEC) of the soil by sustaining cation (sodium, pH, phosphorous, magnesium, calcium) nutrient-holding capacity. Water-holding capacity dictates the length of time between irrigation events, which are 6 hours for the MFCS as per the landscape contractor irrigation schedule. Additionally, precipitation (including sprinklers and drip irrigation) and soil infiltration rates jointly determine the maximum duration of individual irrigation cycles [34]. The duration was established at 18 minutes by the landscape contractor based on the plants and trees' watering need. This finding will help confirm if soil infiltration is affected by the water quality or other factors such as soil structure, degree of compaction, organic matter content, or chemical make-up [34, 53].

3.2 PES 2016 pilot empirical study versus SEI 2017 intervention results

3.2.1 Soil quality test assessment in 2016 against 2017

As discussed in Section 3.1 above, water quality can influence soil quality and its ability to respond to nutrients. This section provides the soil quality results in 2017 against 2016, after solution B application.

In June 2017, a total of six soil specimens were sampled to establish a soil quality baseline against the 2016 pilot project soil test results (see Section 2.3 above).

The results of the pilot project (PES 2016) have been evaluated according to the recommended soil parameters [45–47] to establish why and how the soil conditioner could help reduce water demand for LI. An independent accredited laboratory was employed by the author to conduct the soil test in March 2016 and June 2017. The main results for 2016 and 2017 sampling tests results are summarized in **Tables 3** and **4**.

Following the soil amendment to the whole site in 2017 with Solution B, the pH results in 2017 for samples one, two, and three provide evidence that the soil is neutral [44]. Most micronutrients tend to be less available when soil pH is above 7.5 [47], which is not the case here. However, sample four, five and six results provide evidence the pH is still slightly alkaline when following Flynn's [44] pH concentration limits, although less so than in 2016.

Table 3 provides evidence that in 2017 nitrogen (N) was low for samples one, three, four, and five and deficient for sample two. The potassium (K) level was at moderate level in samples one, four, and five, and low in samples two, three, and six.

In 2017, the organic matter was moderately high and more uniformly established between all samples than in 2016. The organic matter content can help estimate how much N has been supplied to a plant during the growing season [44]. Each percentage of organic matter credits the plant with N (ibid). This is evidenced in **Table 3**. In 2016, the N level was higher than in 2017 and so was the organic matter. If excessive organic matter is added to the soil, infiltration rates can decline [34]. The 2017 soil test results (**Table 3**) provide evidence that the organic matter in the soil was more linear than in 2016 because the landscape contractor uniformly applied the soil conditioner to the landscape site, which may have balanced N in the soil and therefore strengthened the soil texture. Improving the texture of the soil increases the water-holding capacity. There is generally a strong relationship between soil organic matter, soil texture, and water-holding capacity [34, 35].

An adequate level of soil moisture content helps avoids excessively high soil electrical conductivity (EC) and soil salt concentration [34], as evidenced in **Table 4**; moisture content was more uniformly present in 2017 than in 2016 (**Table 3**). Thus, soil conditioner B improved the soil infiltration rate by balancing the soil salt content with other minerals such as magnesium, calcium, and potassium—which means that in 2016 the soil may have had excessive salt content—decreased the infiltration rate and affected the soil moisture content.

In 2017, the exchangeable sodium percentage (ESP) level in all samples was lower than in 2016, except for samples three and four (see **Table 4**). Yet both soil tests showed a level higher than the recommended limit of 15%, which reveals poor soil infiltration [44, 54]. This may be caused by the quality of irrigation water [35]. However, the ESP limits are not fixed values that are to be rigidly applied to all soils. Sandy soil will tolerate much higher ESP values than clay soil [44, 54] as was the case at the MFCS in 2017.

In 2017, the sodium level is higher than the calcium, magnesium, and potassium levels in samples one and two (see **Table 3**). Samples three, four, and five show more calcium, but sodium exceeds magnesium and potassium in all samples except sample five, which has a slightly higher level of magnesium than sodium and potassium.

In relation to **Table 3**, in 2016 the calcium and magnesium were particularly high in sample three, and potassium was lower than calcium and magnesium in all samples except in sample two. Calcium, magnesium, and sodium concentration limits are different for each plant [23]. The concentration limits for the *Pennisetum setaseum* are not known, but it is common for soil in arid regions to have a high calcium level, as was the case here in 2016 [44]. A soil with too much sodium relative to calcium and magnesium is prone to develop problems with water infiltration (ibid), which was the case in 2017 for samples one and two (**Table 3**). Samples two, three, four, and six (**Table 3**) have a higher level of sodium than magnesium. A low-salinity water, like condensate water, will decrease infiltration [35].

In 2016, sample five provided evidence the soil was sodic, while in 2017 the soil was neither sodic nor saline.

The higher the sodium adsorption ratio (SAR), the more likely water will not infiltrate into the soil. This also depends on the irrigation water salinity [44].

Table 3 results provide evidence that in 2017 the SAR ranges were from 0.57 meq/l (sample three) to 1.13 meq/l (sample four). The EC decreased from 2016 to 2017, ranging from 0.18 meq/l (sample two) to 0.74 meq/l (sample four) in the latter year. According to the results of SAR at the EC observed level [35], the soil shows some infiltration problems.

A soil is classified as sodic when the SAR is above 13, the ESP is above 15%, the EC is less than 2dS/m, and the soil pH is greater than 8.5 [47]. The results show that in 2017, the MFCS soil had a neutral pH, a low SAR, but a high ESP and a low EC. This means that the salinity may be affected by the EC of the irrigation water and so the soil may need to be carefully managed due to its low EC conditions (ibid).

Another factor that might be affecting infiltration is the high evaporation rate of AD, particularly in summer, because soil texture plays an important role in the evaporation process [55]. For instance, fine-textured soils have a stronger capillary action, promoting the evaporation of subsurface soil water and bringing salts to the surface, suppressing the osmotically driven water uptake of plants [55]. Therefore, improving soil texture must be part of the solution, as was observed in July 2017 by visual inspection of the MFCS landscape after the soil was enhanced with Solution B (SEI 2017 Intervention).

The 2017 soil test results demonstrate that balancing pH and salinity levels is important to ensure the implementation of the soil enhancement with soil conditioner B is effective and assessable for water conservation reduction and plant growth optimization. The soil amendment brought the CEC to an acceptable level by sustaining cation (sodium, pH, phosphorous, magnesium, calcium) nutrient-holding capacity. Water-holding capacity dictates the length of time between irrigation events. And precipitation (including rain, sprinkler, drip irrigation) and soil infiltration rates together determine the maximum duration of individual irrigation cycles [34]. This is an important finding, which provides evidence that improving the soil structure also increases the soil's water-holding capacity [34].

3.2.2 SEI 2017 intervention soil conditioner results

One month after the soil B application (June 2017), the irrigation controller was programmed to reduce water by 50%, as per the pilot empirical study two results for the whole landscape. Considering the site of the landscape, it took approximately 2 months (mid-March 2017 to mid-May 2017) to apply solution B to the remaining 24,402 m^2 of the MFCS landscape.

From April 2017 through to September 2017 (6 months), plant (*P. setaceum*) growth was monitored though visual observations and weekly photographic reports for each valve location. The 2017 water consumption was analyzed and reported through the EMCS.

Water-holding capacity and infiltration rates are ultimately linked [34]. Solution B enhanced the water-holding capacity of the soil, resulting in better infiltration of the water into the soil because it improved CEC, as evidenced from the 2017 photographs (**Figure 2**) and the soil test results in 2017 as discussed in Section 3.2.1 above.

Following the 2017 soil test results, it was also found that the addition of gypsum, acting as cation nutrient-holding capacity, was needed to maintain the pH level at neutral (see samples four, five, and six, **Table 3**), and increased calcium in the soil was needed to counteract the higher sodium level (see samples one, two, and four, **Table 3**). Gypsum can help correct soil alkalinity caused by a high ESP [35]. The volume of gypsum to apply is determined by the ESP and base saturation percentage sodium values [34]. Amendments comprising soluble calcium salts or acids, or acid-forming substances will be beneficial [56, 57].

In September 2017, the landscape contractor applied a gypsum solution to balance anions and cations, so that pH and salinity levels are adjusted, and the applied soil conditioner is effective for long-term water conservation and plant growth optimization.

The gypsum solution was applied to the surface of the landscape with the addition of light hand-watering over the surface to settle the gypsum down and avoid the

Figure 2.
SEI 2017 Results—Plant Growth Solution A (left) and B (right).

Total valves	A	B	C	Total Water loss based on 16 hours irrigation time per day (m^3)
	Water loss (liters) every 18mns	**Under irrigated plants (liters) every 18mns**	**Difference in liters (A-B)**	
		Controller A		
52	32.46	−5.19	27.27	1.45
		Controller B		
83	60.86	−7.58	53.28	2.84
		Total controller A and B		
135	93.32	−12.77	80.55	4.30
Total water loss per year in m^3				1567.94

Table 5.
SEI 2017 intervention outdoor valve flow audit results (primary data collected by the third-party auditor and analyzed by the main author).

solution B soil conditioner washing away [49]. For this reason, a water amendment, such as chemigation, may be more appropriate for the MFCS to save on manpower cost and time of application [34].

3.2.3 SEI 2017 outdoor valve flow audit results

As discussed in Sections 2.2 and 3.1, the remaining 127 outdoor valve flow was verified (audit initiated by the main author) to minimize further non-potable water loss. Following the application of Solution B for the entire site, the irrigation controller had to be reprogrammed to the new irrigation rate based on ADM standard [28] with 50% water reduction. The reprogramming of the irrigation controller with the revised flow rate in liters per second should help save an additional 1567 m^3 of water per year (see **Table 5**).

Table 5 below shows the results of the 135 valves audited at the site, which provides evidence the MFCS wasted 1567 m^3 of water per year. Most of the valves had a higher flow rate than the As-Built (April 2015 hand-over documents to building operator) documentation records, resulting in the landscape being over-watered by 4 m^3 per day.

4. Discussion

4.1 Revised irrigation rate standards against PPC1 calculation

As discussed in Section 2.3 above, the UPC irrigation rate [30] has been revised since 2010. In 2016, the irrigation rate (UPC) [29] was deemed impossible to apply by the landscape contractor, who feared plants would not survive watering at 2.5-day intervals in the harsh climate, particularly during summer (42.6°C mean maximum temperature during the day). The revised 2017 rate calls for a three-day irrigation interval in summer and 7 to 10 days in winter. Thus, the revised UPC irrigation rate [30] would still be challenging to apply due to high evaporation rate and higher plant water need in summer; the recommended volume of 107 m^3/day is deemed very low (**Figure 3**). This has been evidenced during the pilot conditioner testing (PES 2016). When the LI was reduced to 60% (equivalent to 166 m^3/day) in July 2016, the *Pennisetum setaseum* showed signs of water stress.

It was found that the MFCS irrigation rate after soil improvement was 50% less than the ADM [28] irrigation rate, 35% less than the UPC [30] guidance and 5% above the U.S. EPA LI budget calculation [31]. In summer, that equates to 416 m^3/day [28] against 208 m^3/day (SEI 2017); 321 m^3/day [30] against 208 m^3/day (SEI 2017) and 208 m^3/day (SEI2017) against 198 m^3/day [31], respectively. That means the ADM [28] and the UPC [30] standards in their current state are erroneous, because they encourage landscape professionals to either wastewater by over-irrigating [28] or under-irrigating [30] in the case of the revised irrigation rate (**Figure 3**) based on three-day intervals [30], as discussed with the landscape contractor.

Additionally, it was found that the MOCCAE soil standard [27] does not clearly indicate minimum and maximum soil micro- and macronutrient concentration limits

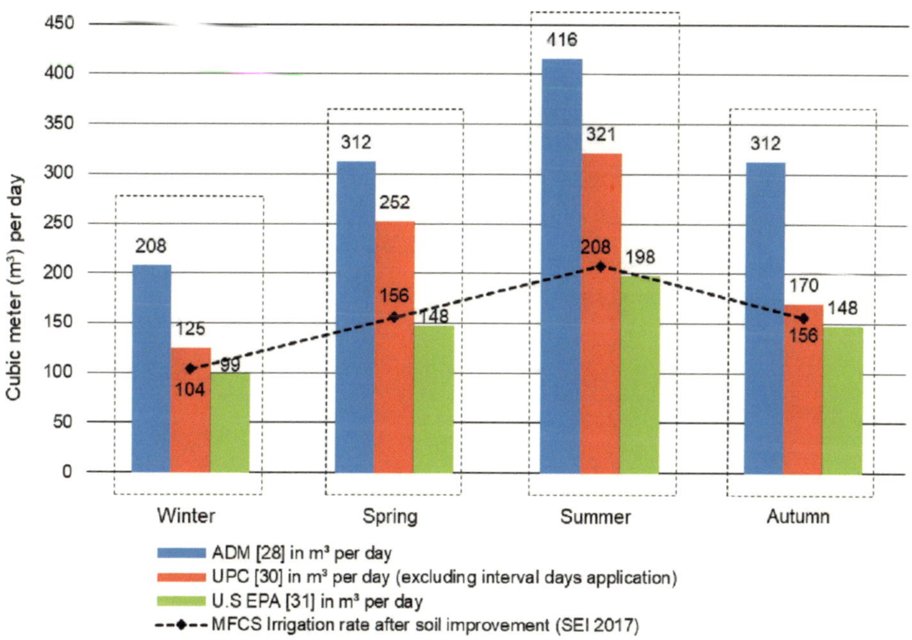

Figure 3.
SEI 2017 Irrigation Rate against ADM [28], UPC [30], and U.S. EPA [31].

for soil maintenance. Moreover, the RSB standard [26] does not include salinity water parameters, which could be beneficial to irrigation water. These two aspects point to the importance of direction from the authority to direct operations and maintenance teams on outdoor water use water and soil quality requirements for UAE climatic conditions, and to promote techniques to increase soil water-holding capacity to prevent soil deficiency and soil infiltration problems.

4.2 Water demand outcome

Calculation one (Calc1 2017) is as per PPC1 2016 calculation one, which aligns with U.S. EPA [31] water budget with 5% adjustment and 50% less than the ADM standard [28] after soil improvement.

In 2016, the LI condensate water deficit was established at 19,235 m^3 against 11,189 m^3 in 2017, predominantly occurring in the winter and spring months, representing a 42% reduction between the 2 years. In 2016, the deficit occurred for 6 months from December to May and in 2017 for 7 months from November to May.

Figure 4 provides a monthly breakdown of the LI consumption for the year 2017 as recorded by the EMCS, against the LI demand based on ADM standard [28].

In addition, and as shown in **Figure 4**, the water consumption does not follow the pattern of the actual water demand of the plants, shrubs, and trees based on the ADM standard irrigation rate [28] before soil amendment. In March, April, and September, more water was consumed than required according to the standard (ibid.). In

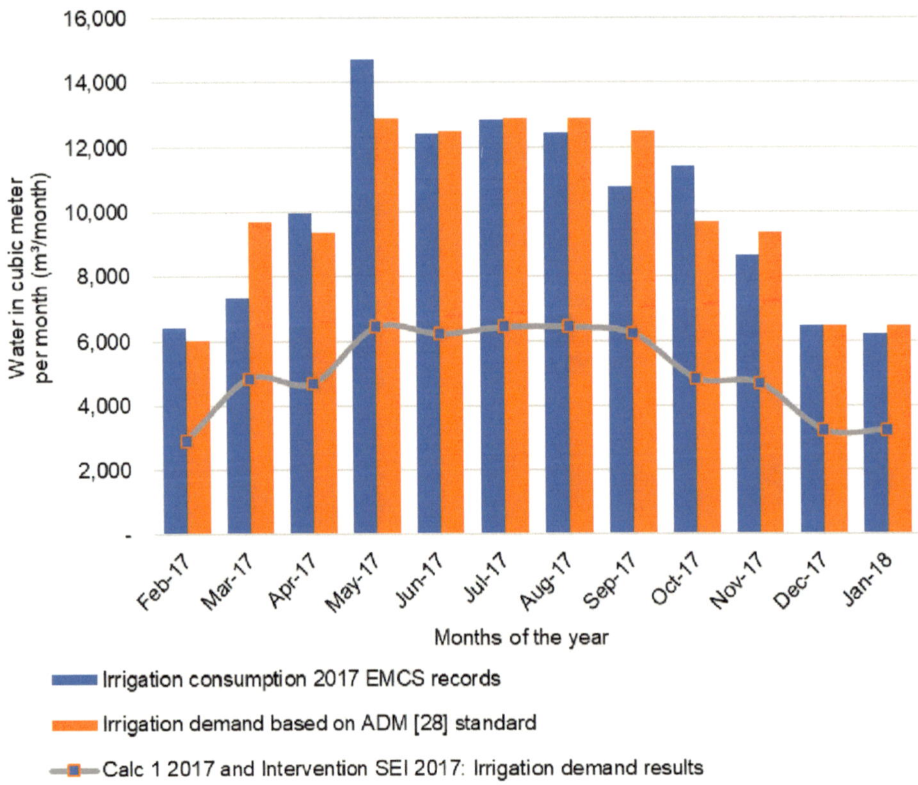

Figure 4.
LI consumption based on 2017 EMCS records against SEI 2017, PPC1 2016, and calculation one (Calc1) results.

addition, from July—the time at which the soil amendment dictated the irrigation rate based on CS1 Calc 1, or 50% less water than the ADM standard [28]—the irrigation was not changed to this new pattern of water demand. Thus, from July through to January 35,100 m^3 of water should have been used for LI, when in fact 65,274 m^3 (an additional 46%) was consumed.

The reason for this excessive water consumption may be either that the landscape contractor did not reprogram the irrigation controller after the 2017 valve flow audit to align with the soil enhancement irrigation rate (SEI 2017 and Calc1 2017), or that there is a leakage outdoors. As of April 2018, the LI and the WFs incurred a condensate water deficit for 7 months of the year. This data analysis provides evidence that the consumption is above the required demand.

The above observations provide evidence of the condensate water deficit for 2017 based on the 2017 EMCS records. From November through to May, there is insufficient condensate water (−11,189 m^3) to feed the 2017 excessive consumption of the LI.

5. Conclusions

The case study presented describes two interventions initiated from April 2016 to September 2017 to increase the water-holding capacity of the soil and thereby save water. Three solutions were tested onsite for 12 months, from March 2016 to March 2017. Soil conditioner B was applied to the remainder of the landscape in April and May 2017, and its impact on water-holding capacity took effect from May to June 2017. A soil test was conducted in June 2017 to evaluate the soil enhancement. The landscape irrigation (LI) valve flow audit was completed in August 2017 for the 127 remaining valves. The reprogramming of the irrigation controller, which aligns with the latter two tasks, was implemented in September 2017. The methodology for soil quality testing, soil conditioner, and valve flow audit application, and irrigation rate calculation have also been discussed.

The findings of the SEI 2017 intervention are threefold. First, the 2017 soil test results provide evidence that the application of gypsum acting as additional cation nutrient-holding capacity was needed in addition to solution B to maintain the pH level at neutral and increase calcium in the soil to counteract the higher sodium level. Second, the MFCS irrigation rate after soil improvement is 50% below the ADM [28] irrigation rate, 35% less than UPC [30] in summer without interval days irrigation, and 5% above the U.S. EPA LI budget calculation [31]. Third, the outdoor valve flow audit helped save 1576 m^3 of water per year during operations and contributing to the reprogramming of the irrigation controller together with an appropriate irrigation rate. The combination of these strategies' implementation helped reduce the A/C condensate water deficit by 42% (−8046 m^3) in 2017 against 2016.

These results provide evidence that, firstly, the ADM [28] standard is applicable only if the soil is amended and the irrigation rate adjusted as per the above method. It was also observed in July 2016 that the UPC standard [29] 2.5-day irrigation interval pattern in summer resulted in plant water stress. Secondly, soils in desert type climates must be amended with both a soil conditioner able to alternate the water-holding capacity of the soil to achieve up to 50% LI reduction and a gypsum solution to balance the pH and anions, to avoid infiltration problems. Thirdly, the most stringent irrigation rate can be applied to a sandy soil in the climatic context of the MFCS only if soil is enhanced to sustain cation nutrient-holding capacity and thereby increase water-holding capacity. This was observed visually in 2016 when solution B was compared with solutions A and C and confirmed in 2017.

These elements point to the importance of direction from the relevant authority to direct operations and maintenance teams on outdoor water use and soil quality requirements for UAE climatic conditions, and to promote techniques to increase soil water-holding capacity to prevent soil deficiency and soil infiltration problems.

This research contributes to the existing body of knowledge in practice by demonstrating that outdoor water demand management has a large role to play in helping minimize water wastage, and thereby decrease GHG emissions, for a medical facility in AD. In addition to this, the authors have demonstrated that a water demand strategy can be used as a decarbonization strategy in AD and may also be transferred to other countries located in arid climates that have limited access to natural resources and depend on desalinated water. The authors, therefore, encourage the UPC [30], ADM [28], and RSB [26] to amend their guidelines and standards by adopting the following recommendations:

- A soil standard should be made available to the public to reflect AD soil and climate conditions.

- The use of soil enhancement techniques for large landscape sites at the design, construction, and operation stage should be mandated to save on water consumption for LI.

- The UPC [30] and ADM [28] irrigation rate should align to the U.S. EPA [31, 32] standard and include the application of soil enhancement as a requirement.

- The commissioning and ongoing commissioning of new buildings' water systems should be mandated to avoid faulty watering rates.

- The RSB [26] recycled water regulations should include salinity parameters for non-potable water to avoid soil infiltration problems.

Other recommendations from this chapter are to:

- Confirm if soil infiltration is affected by water quality or other factors such as degree of soil compaction, organic matter content, or water chemical makeup.

- Supplement the measurements with soil moisture sensors in the field at least at three depths.

- Statically evaluate each sample taken of different depths.

- Include the data of fresh and dry weights of specified plants.

- Test water-binding products improving soil properties for future studies.

Conflict of interest

The authors declare no conflict of interest.

DOI: http://dx.doi.org/10.5772/intechopen.107297

Author details

Geraldine Seguela*, John Richard Littlewood and George Karani
Cardiff Metropolitan University, Cardiff, United Kingdom

*Address all correspondence to: seguelag@gmail.com

References

[1] Köppen and Geiger. World maps of Köppen-Geiger climate classification. 2018. Available from: http://koeppen-geiger.vu-wien.ac.at/present.htm. [Accessed: 2017-02-06]

[2] Veerapaneni S, Long B, Freeman S, Bond R. Reducing energy consumption for seawater desalination. Journal of American Water Works Association. 2007;**99**(6):95-106. DOI: 10.1002/j.1551-8833.2007

[3] Seguela G, Littlewood JR, Karani G. A GHG metric methodology to assess onsite buildings non-potable water system for outdoor landscape use. Journal of Applied Sciences. 2020;**10**. DOI: 10.3390/app10041339

[4] Creswell JW, Creswell JD. Research and Design Qualitative, Quantitative and Mixed Methods Approaches. 5th ed. Thousand Oaks: Sage; 2018

[5] Creswell JW, Plano Clark VL. Designing and Conducting Mixed Methods Research. 2nd ed. Los Angeles: Sage; 2011

[6] Yin RK. Case Study Research: Design and Methods. 5th ed. Thousand Oaks: Sage; 2014

[7] IPCC. Intergovernmental Panel on Climate Change. Climate Change 2014: Synthesis Report. United Nations Environment Programme (UNEP) and the World Meteorological Organization (WMO). 2014. Available from: http://www.ipcc.ch/report/ar5/syr/. [Accessed: 2017.02.06]

[8] EAD. Environment Agency-Abu Dhabi. Abu Dhabi State of Environment Report 2017. Available from: https://www.ead.ae/Publications/Abu%20Dhabi%20State%20of%20Environment%20Report%202017/EAD-full-report.pdf. [Accessed: 2018.03.28]

[9] Pizzi NG. Water Treatment: Principles and Practices of Water Supply Operations. 4th ed. Denver, CO, USA: American Water Works Association; 2010

[10] Mitchell K. From increasing supply to managing demand. In: Reisz, T, editor. Al Manakh 2. Cont'd. Amsterdam: Stichting Archis; 2009. p. 114-117. ISBN-10: 9077966234

[11] World Resource Institute. Water risk atlas. 2019. Available from: https://www.wri.org/aqueduct. [Accessed: 2020.12.06]

[12] Verner D. Adaptation to a Changing Climate in the Arab Countries: A Case for Adaptation Governance and Leadership in Building Climate Resilience. World Bank; 2012. DOI: 10.1596/978-0-8213-9459-5

[13] Balfaqih H, Al-Nory MT, Nopiah ZM, Saibani N. Environmental and economic performance assessment of desalination supply chain. Desalination. 2017;**406**:2-9. DOI: 10.1016/j.desal.2016.08.004

[14] Pacia D, Yeh D. Adaptation to climate for water utilities. In: Ahuja S, editor. Water Reclamation and Sustainability. 1st ed. Waltham: Elsevier; 2014. pp. 19-50. DOI: 10.1016/C2012-0-07359-0

[15] Ali SE, Alsaman AS, Harby K, Askalany AA, Refaat Diab M, Ebrahim Yakoot SM. Recycling brine water of reverse osmosis desalination employing adsorption desalination: A theoretical simulation. Desalination. 2017;**428**:13-24.

DOI: 10.1016/j.desal.2016.12.002 0011-9164

[16] Shahid SA, Abdelfattah MA, Othman Y, Kumar A, Taha FK, Kelley JA, et al. Innovative thinking for sustainable use of terrestrial resources in Abu Dhabi Emirate through scientific soil inventory and policy development. In: Shahid S, Taha F, Abdelfattah M, editors. Developments in Soil Classification, Land Use Planning and Policy Implications. 1st ed. Springer: Dordrecht; 2013, 2013. pp. 3-49. DOI: 10.1007/978-94-007-5332-7_1

[17] Environment Agency Abu Dhabi. The Water Resources Management Strategy for the Emirates of AD 2014-2018 [internet] 2014. Available from: https://www.ead.ae/Documents/PDF-Files/Executive-Summary-of-The-Water-Resources-Management-Strategy-for-the-Emirate-of-Abu-Dhabi-2014-2018-Eng.pdf. [Accessed: 2017-02-26]

[18] USDA. United States Department of Agriculture. Assessment of Water Holding Capacity of Soils Map [internet]. 2012. Available from: https://www.nrcs.usda.gov/wps/portal/nrcs/detail/soils/use/?cid=nrcs142p2054022. [Accessed: 2016-12-03]

[19] King P, Grealish G, Shahid SA, Abdelfattah MA. Land evaluation interpretations and decision support systems: Soil survey of AD Emirate. In: Shahid S, Taha F, Abdelfattah M, editors. Developments in Soil Classification, Land Use Planning and Policy Implications. 1st ed. Dordrecht: Springer; 2013. pp. 147-164. DOI: 10.1007/978-94-007-5332-7_1

[20] Ministry of Environment and Water. Ministerial resolution number (476) of the year 2007 Concerning by-law of AGCC fertilizers and agricultural soil conditioners law. 2007. Available from: https://www.moccae.gov.ae/assets/download/e4e8ae38/476.pdf.aspx. [Accessed: 2017-02-17]

[21] Heid J. The paradox of sustainability and Abu Dhabi. In Reisz, T. editor. Al Manakh 2. Cont'd. Amsterdam: Stichting Archis; 2009. pp. 118-119. ISBN-10: 9077966234

[22] Benkhelifa M, Daoud Y, Tessier D. Effects of maghnian bentonite on physical properties of Sandy soils under semi-Arid Mediterranean climate. Pakistan Journal of Biological Sciences. 2008;**11**:17-25. DOI: 10.3923/pjbs.2008.17.25

[23] Jr BJ, J. Plant Nutrition and Soil Fertility Manual. 2nd ed. Boca Raton: CRC Press; 2012. DOI: 10.1201/b11577

[24] IEA. International Energy Agency (2016) Water Energy Nexus. Paris: OECD/IEA [internet] 2016. Available from: https://www.iea.org/publications/freepublications/publication/WorldEnergyOutlook2016ExcerptWaterEnergyNexus.pdf. [Accessed: 2017-03-12]

[25] Ferguson BK. Water conservation methods in urban landscape irrigation: An exploratory overview. Journal of the American Water Resources Association. 2007;**23**:147-152. DOI: 10.1111/j.1752-1688. 1987.tb00794.x

[26] RSB. Regulatory Supervision Bureau. Guide to recycled water and biosolids regulations [internet]. 2010. Available from: http://rsb.gov.ae/assets/documents/264/regsrwb2010.pdf. [Accessed: 2012-09-12]

[27] MOCCAE. Ministry of Climate Change and Environment. Ministerial resolution number (476) of the year 2007 concerning by-law of AGCC fertilizers and agricultural soil conditioners law.

[internet]. 2007. Available from: https://www.moccae.gov.ae/assets/download/e4e8ae38/476.pdf.aspx. [Accessed: 2017-02-17]

[28] ADM. Abu Dhabi Municipality. Irrigation Systems Operation and Maintenance. Parks and Recreation Facilities Division: Abu Dhabi. 2C Section 02800 ed. Abu Dhabi: ADM; 2013. p. 5

[29] UPC. Urban Planning Council. Abu Dhabi Public Realm Design Manual [internet]. 2010. Available from: https://www.cip-icu.ca/pdf/2011-HM-Urban-Design2. [Accessed: 2016-06-02]

[30] UPC. Urban Planning Council. Appendices Public Realm Design Manual. [internet]. 2017. Available from: https://www.upc.gov.ae/en/publications/manuals-and-guidelines/appendices-public-realm-design-manual. [Accessed: 2018-04-18]

[31] U.S. EPA. (United States Environment Protection Agency). Watersense landscape water budget tool Version 1.02 [internet]. 2014. Available from: https://www3.epa.gov/watersense/water budget. [Accessed: 2017-01-03]

[32] U.S. EPA. (United States Environment Protection Agency). WaterSense water budget approach [internet]. 2014. Available from: https://www3.epa.gov/watersense/docs/home_final_waterbudget508.pdf. [Accessed: 2017-01-03]

[33] Seguela G, Littlewood JR, Karani G. Eco-engineering strategies for soil restoration and water conservation: Investigating the application of soil improvements in a semi-arid climate in a medical facility case study, Abu Dhabi. Journal of Ecological Engineering. 2017;**121**:53-64. DOI: 10.1016/j.ecoleng.2017.07.020

[34] Duncan RR, Carrow RN, Huck MT. Turfgrass and Landscape Irrigation Water Quality: Assessment and Management. Boca Raton: CRC Press; 2009

[35] Ayers RS and DW Westcot. Water Quality for Agriculture. Irrigation and Drainage Paper, 29, Rev. 1. Food and Agric. Organization of the United Nations [internet]. 1994. Available from: http://www.fao.org/DOCREP/003/T0234E/T0234E00.htm#TOC. [Accessed: 2017-02-17]

[36] Lee J Landscape Plants of Arabia. London: Gilgamesh-Publishing; 2013. ISBN-10: 1908531037

[37] ADA. Arriyadh Development Authority. Manual of Arriyah plants [internet] 2014. Available from: http://www.ada.gov.sa/idc/groups/public/documents/AR_ADA_Prints/006792.pdf. [Accessed: 2016-02-17]

[38] DMAT. Department of Municipal Affairs and Transport. Abu Dhabi International building code: Plumbing systems. Chapter 29. p. 501 [internet]. 2013. Available from: https://www.yumpu.com/en/document/view/51842029/the-code-handbook-abu-dhabi-international-building-codes. [Accessed: 2017-05-02]

[39] Wilchelns D, Qadir M. Achieving sustainable irrigation requires effective management of salts, soil salinity, and shallow groundwater. Agricultural Water Management. 2015;**157**:31-38. DOI: 10.1016/j.agwat.2014.08.016

[40] Fay M, Hallegatte S, Vogt-Schilb A, Rozenberg R, Narloch U, Kerr T. Decarbonizing Development: Three Steps to a Zero-Carbon Future. Climate Change and Development; Washington, DC: World Bank [internet]. 2015. Available from: http://www.worldbank.

org/content/dam/Worldbank/document/Climate/dd/decarbonizing-development-report.pdf. [Accessed: 2018-03-18]

[41] Lado M, Ben-Hur M. Treated domestic sewage irrigation effects on soil hydraulic properties in arid and semi-arid zones: A review. Soil and Tillage Research. 2009;**106**:152-163. DOI: 10.1016/j.still.2009.04.011

[42] Bedbabis S, Ben Rouina B, Boukhris M, Ferrara G. Effect of irrigation with treated wastewater on soil chemical properties and infiltration rate. Journal of Environmental Management. 2014;**133**:45-50. DOI: 10.1016/j.jenvman.2013.11.007

[43] Britton Harrell J. An evaluation of soil sampling methods in support of precision agriculture in Northeastern North Carolina [thesis]. University of Southern California; 2014

[44] Flynn R. Interpreting soil tests: Unlock the secrets of your soil. Las Cruces, USA: New Mexico State University, College of Agricultural, consumer and environmental sciences. [internet] 2015. Available from: https://pubs.nmsu.edu/_circulars/CR676/. [Accessed: 2017-09-20]

[45] Hornek DA, Sullivan DM, Owen JS, Hart JM. Soil Test Interpretation Guide. Oregon State University. 2011. [internet]. Available from: http://extension.oregonstate.edu/sorec/sites/default/files/soil_test_interpretation_ec1478.pdf. [Accessed: 2018-09-28]

[46] Duncan RR, Carrow RN, HuckM T. Treating the cause, not the symptoms: Irrigation water treatment for better infiltration. United States Golf Association (USGA) Green Section Record [internet]. 1999;**37**(6): 11-15. Available from: http://gsr.lib.msu.edu/1990s/1999/991111.pdf. [Accessed: 2016-03-04]

[47] Jensen T. Soil pH and the availability of plant nutrients. In: Plant Nutrition Today [internet]. 2010; Vol. 2: 2. Norcross: International Plant Nutrition Institute (IPNI). Available from: www.ipni.et/publication/pnt-na.nsf. [Accessed: 2016-03-04]

[48] UPC. Urban Planning Council. Pearl building rating system: Design & construction, Version 1.0. [internet] 2010. Available from: https://www.upc.gov.ae/en/publications/manuals-and-guidelines/public-realm-rating-system. [Accessed: 2010-10-15]

[49] Bondzic M. NRM laboratories, Nutrient Solution. 2018. [by email]

[50] Heydon B. Open keynote: An update on Abu Dhabi City's Municipality's new public realm guidelines. In: Future Landscape & Public Realm Conference. Abu Dhabi. 2017

[51] Farley M, Trow S. Losses in Water Distribution Networks: A Practitioners' Guide to Assessment, Monitoring and Control. 2003. London: IWWA Publishing. eISBN: 9781780402642

[52] Hamilton S, McKenzie R. Water Management and Water Loss. London: IWWA Publishing; 2014. eISBN: 9781780406367

[53] Duncan RR, Carrow RN, Huck, M T. Understanding water quality and guidelines to management. United States Golf Association (USGA) Green Section Record. [internet] 2000;**38**(5):14-24. Available from: http://gsrpdf.lib.msu.edu/ticpdf.py?file=/2000s/2000/000914.pdf. [Accessed: 2016-03-04]

[54] Flynn R, Ulery A. An Introduction to Soil Salinity and Sodium Issues

in New Mexico. Las Cruces, USA: New Mexico State University. College of Agricultural, Consumer and Environmental Sciences. 2011 [internet]. Available from: http://aces.nmsu.edu/pubs/_circulars/CR656.pdf. [Accessed: 2017-09-20]

[55] Laity J. Deserts and Desert Environments. 2008. West Sussex: Wiley-Blackwell. ISBN: 978-1-577-18033-3

[56] FIFA. Fertilizer Industry Federation of Australia Inc. In: Price G, editor. Australian Soil Fertility Manual. 4th ed. Canberra: FIFA; 2006

[57] Abrol IP, Yadav JSP, Massoud FI. Salt-Affected Soils and their Management. FOA Publishing. Rome: Food and Agriculture Organization of the United Nations [internet] 1998. Available from: http://www.fao.org/docrep/x5871e/x5871e05.htm. [Accessed: 2016-02-17]

Chapter 11

Global Water Mapping, Requirements, and Concerns over Water Quality Shortages

Aijaz Panhwar, Rashid Abro, Aftab Kandhro, Abdul Rauf Khaskheli, Nusrat Jalbani, Khadim Ali Gishkori, Atta Muhammad Mahar and Sofia Qaisar

Abstract

Water is a blessing and reason forsurvival on earth for human beings. Without water, it is not difficult to live on earth, but it is impossible. Water fulfills domestic, drinking, agricultural, and industrial requirements and makes everything useful for us. Unfortunately, rapid increase in population, unplanned agriculture practices, and industrial sector have become a serious issues for environment and for future needs. Water will be a very important issue for next generations to face. Safe water access up to 50–100 liters per day for drinking and sanitation is a right of every human being, recognized by the United Nations General Assembly in July 2010. Approximately, six billion people may suffer from scarcity of clean water by 2050. The agriculture sector is the largest consumer of freshwater around 70% followed by industry and domestic of 19 and 11%, respectively. The global demand for water is gradually increasing by 1% per year. Global demand for water is predicted to increase by 55% between the years 2000 and 2050. Existing global water demand has been projected to about 4600 km^3 every year and estimation may increase up to 20–30%, in range of 5500–6000 km^3 every year by 2050.

Keywords: agriculture, crops, environment, fertilizers, global, irrigation, population, pollution, sanitation, wastewater treatment

1. Introduction

Water is life; without water, survival is not possible. No water, no survival on earth. The safe water is right of all human beings on earth, to provide 50–100 liters of water for each person/day, for use in domestic and other necessities. It is fact that water is a precious gift on earth for us; 4000 km^3 of water withdrawal is done by human each year at rate of 1.6% of groundwater deposits. For unplanned increase in population, it is expected that water demand may increase up to 55%, 2000–2050. It will put more pressure on groundwater resources. The reason of the demand/ pressure on groundwater resource is rapid urbanization, industrialization, and old practices in agriculture sector. This means that millions of vulnerable families around the world do not drink, cook, or bathe with clean water. Groundwater is a major and important source for drinking purpose, irrigation, and industrial sectors. Major portion of groundwater around 65% is used as a drinking water, 20% for irrigation and livestock (agriculture), and 15% for industrial sector. The majority of the population depends on groundwater source. Groundwater resources are severely affected by unplanned increase in urbanization and traditional irrigation practices. This act does not decrease quantity of groundwater only, but it is also deteriorating its quality. The requirement of a person is 2–4 liters/day drinking water and up to 200 liters for sanitation and other necessities as described in **Figure 1**. Safe drinking water has become a dream for millions of people on the earth. Millions of populations have no access to safe drinking water and sanitation and are living without substantial supply of water. Around 4.5 billion (61%) have no access to proper sanitation and 2.1 billion (29%) no access to safe water [1]. Around minimum 200 liters of water is needed for feeding of one person per day [2]. Burek et al. [3] have estimated the requirement of water in different agriculture crops water needed by 2050, up to 23–42% above the level in 2010 [4]. Agriculture/irrigation is the biggest consumer of freshwater, as its share is 70%, and **Figure 2** describes the sector wise use of freshwater. As per report, water demand is gradually increasing at 1%/year. Scarcity of clean water has become serious issue of entire world population. If this increase of population pressure continues, it may reach up to 9.4–10.2 billion around 22–34% increase. Almost 50% of the population of entire world is living in mega cities called urban areas. The ratio of migration from rural to urban is more than expectations. The reason of migration is less facility of jobs, clean water, sanitation, poor education, and health system. Due to more facilities, the life of urban areas is easier as compared to rural area life. Presently, around 47% of the population is living in urban areas. Rapid migration to urban areas seems by 2050 around 52–65% of the population will be shifted in urban areas; every 07 out of 10 or 2/3 of the population may live in urban areas [5]. By 2050, the water quality and availability depend upon increase of population migration, industrial development, living standard, and quality of life. Around 57% of population may suffer for scarcity of water for 1 month in each year. In developing countries, this issue still only discusses water quantity and quality but no control. The situation is deteriorating and will be difficult/hard to control over the situation [6]. The consumption of water in United States of America has decreased dramatically in all sectors such as Domestic/Municipal, Industrial as well as agriculture [7]. The General Assembly by

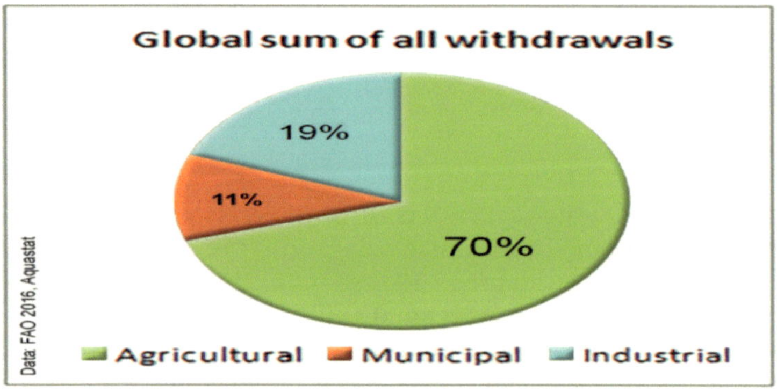

Figure 1.
Basic water requirements (per person) for human needs.

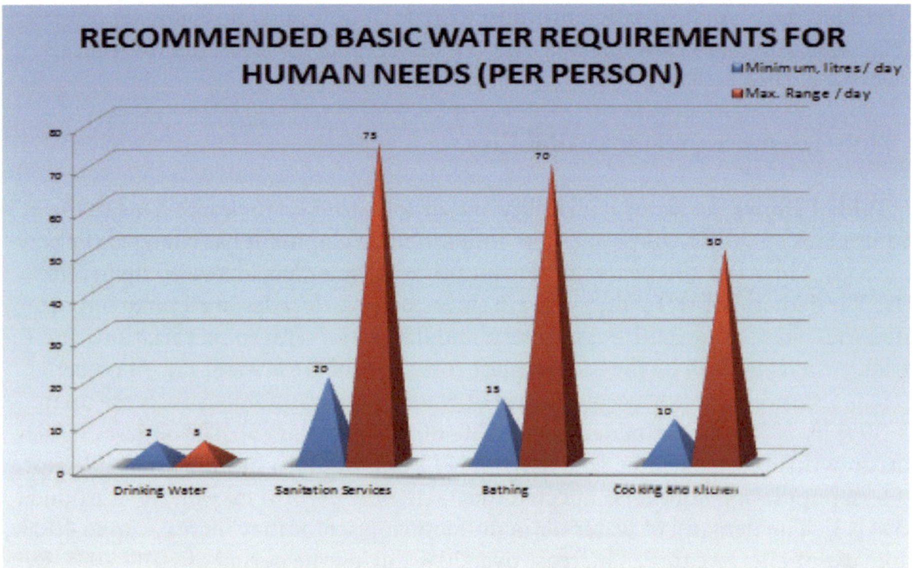

Figure 2.
Global sum of all withdrawals.

the UN recognized a historical decision for access to water and sanitation. The water costs should be within 3% of the income. The safe water should be in access, not away more than 1000 meters or within 30-minute distance from home. In 2016, India was the highest user of groundwater and the usage was more than China and even United States America [8]. The draw of groundwater in India 109–245 m^3, and the major share was used in agriculture sector; around 90% public of rural areas are used for house as well as for drinking purpose. In the last 50 years, the human population has more than doubled. This rapid growth with its accompanying economic development and industrialization has transformed water ecosystems around the world and resulted in a massive loss of biodiversity [9]. Nearly 50% of the world population is already living in potential water scarce areas and the number may increase for 4.8–5.7 billion in 2050. Almost 73% of the affected people are living in Asia (69% by 2050). As of 2015, 2.1 billion people (29 percent of global population) lacked access to safely managed drinking water and 4.5 billion people (61%) lacked properly managed sanitation. Due to advancement of technology, excess use of fertilizers is the main source of deterioration of water quality and has created serious concerns for environment, such as our drinking water containing various impurities of different nature such as biological, chemical, and physical. The utmost hazardous contamination is of biological nature, which causes serious human health problems and may cause death in many cases. About 10 billion tons of freshwater used by human result in severe pollution return to environment. It is estimated that about 80% of wastewater discharged from industries and municipal is directly released to the environment without any prior treatment, subsequently in a growing deterioration of overall water quality with the detrimental impacts on human health and ecosystems. Throughout world about 2.2 million people die every year by diarrheal disease. In developing countries, the main cause of 80% diseases is water related. The high-income countries treat about 70% of wastewater generated by industrial and municipal wastewater. The ratio of treatment is different for upper middle-income countries, and 38% of generated wastewater is treated but very worst situation prevails in lower middle-income countries of up to 28%, while only 8% in low-income countries undergoes for treatment for reuse.

2. Global drinking water availability

Table 1 shows the water availability and distribution on the earth. Due to the rapid and unplanned growth of population, industrial development has changed the way of use [5]. Since last 100 years the consumption of water has increased up to 600% [11]. The demand of water may grow in the next three decades in all three important categories, such as agriculture, industry, and domestic, with some variations and the agriculture sector will be the same largest consumer of freshwater [5]. At present, there is around 4600 km^3/year demand of water for use and it may increase up to 20–30% by 2050. It is estimated up to 5500–6000 km^3/year [3]. The largest population growth/increase will be seen in Africa (13,000 million) or more than 100% of present population [12]. It is expected that GDP of the world may increase around 2.5% [13]. The demand of water for manufacturing sector may increase up to 400%, while use of energy can also increase over 20% during the period of 2010–2035. By 2050, the North America and Western Europe will be the largest user of the industrial water. This increase will go up to 800% in African continent. While at present there is nominal use of water in this continent, other side is of about 250% in Asia continent [11]. The use of domestic water scenario is different out of 100, which will be 10%. Definitely it will increase as population increases. The situation will be more different by 2050 in Asian and African countries; around 300% demands will increase, while in South and Central America up to 200% increase. The demand can increase dramatically throughout the continent and will be difficult to handle.

Water distribution	Percentage (%)
Rivers	0.0001
Atmosphere	0.001
Soil moisture	0.005
Inland seas	0.008
Freshwater lakes	0.009
Groundwater	0.61
Glaciers and other ice	2.15
Ocean water	97.2
Percentage earth's surface covered by water.	71
Available water on the planet (Cubic Miles)	326 million
Freshwater on earth is unavailable: protected up in glaciers, atmosphere, polar ice caps, and soil; either it is highly polluted; or rests too far off the earth's surface and to be extracted at an affordable cost.	2.5
Availability of freshwater on surface of earth	0.5

Table 1.
Distribution of the water on earth [10].

3. Global water demand

Figure 2 shows that agriculture is the largest consumer of freshwater uses, and there is around 70% of the world's accessible freshwater, but some 60% of this is wasted due to leaky irrigation systems. Agriculture and energy sectors are very important for human and both are water intensive. It is estimated the increase around 60% and 80% in both sectors [14, 15]. The rivers and lakes are in worst condition and almost polluted in Asia, Latin America, and Africa. Due to industrialization, climate change may change/increase in pollution level, which is a great threat to environment, aquatic system, human health, and hurdle in sustainable development goals [16]. The major reason of water quality deterioration is loading of nutrient and pathogen. There is also great impact of chemicals on water quality. The pollutant will increase more in lower- and lower middle-income countries, due to rapid and unplanned increase in population, uncontrolled economic activities, and lack of funds as well as the management of wastewater [16]. The withdrawal at present is at the highest level, of around 4600 km^3/year [17, 18]. Presently around 800 million people are suffering from food; by 2050 the need/demand of food production will be increased 50% for the feeding of more than 09 billion population, estimated on earth [19]. Around 800 km^3/year groundwater is used for agricultural irrigation sector in 2010. India was the largest consumer followed by USA, China, Iran, and Pakistan. These countries are using 67% of the total groundwater [3]. It is projected that groundwater withdrawal may increase more 39% amounting to 1100 km^3/year by 2050.

4. Water scarcity

As population is increasing, the use of water also increases. This situation creates shrinking position of the water resource availability. The drinking water quality is already affected by industrial development due to physical, biological, and chemical impurities. Among these, some impurities such as biological may lead to death [20]. Many countries are facing water shortage conditions and will face more wore worst conditions in reduction of water availability by 2050 [21]. In 2010, 27% or 1.9 billion of the population was living in water shortage areas. The population ratio may go up to 42–95% or 2.7 billion of the population. The majority of affected water shortage areas are living in Asia, and the ratio is around 73% of the global population [5]. Groundwater usage has become common without groundwater reality for its depletion [22]. Presently 3.6 billion people around 50% of the world's population is residing in water scarcity areas, which may increase up to 33–58% with 4.8–5.7 billion population by 2050 [21]. Almost 1100 million people around the world have lack of access to water; approximately 2700 million have water scarcity issue at least for a month in the whole year. Over a 2400 million people are exposed to diseases such as cholera, typhoid fever, and other water-borne diseases due to the improper sanitation. Considering 02 million people most of them are children per year and die from diarrheal diseases alone. About 2/3rd of the world's population may face the water scarcity problem by 2025 as per the current consumption rate, making the ecosystem of the world to suffer more. Inadequate access to freshwater already influences the development opportunities and lifestyle in water scarce areas [23]. Water (per capita) consumption changes considerably around the world. Countries or regions that are already developed may consume an average of 200 L/per person/per day.

Internationally adopted value for consumption of water is around 50 L/per person/ per day, shown in **Figure 1** [17]. Recently, a report was published on USA's water consumption reduction in almost all sectors, municipal, agricultural, industrial, and thermoelectric power, but the report concludes that while substantial progress has been made, current water use trends are not sustainable in the face of population growth and climate change [7]. By the year 2025, as much as 60% of the global population may suffer from physical water scarcity [24, 25].

5. Water pollution

The treatment and wastewater management is reflex of the country's commitment, development, and its income level. Water pollution comes from many sources including pesticides and fertilizers that wash away from farms, untreated human wastewater, and industrial waste. Even groundwater is not safe from pollution, as many pollutants can leach into underground aquifers. Some effects are immediate, as when harmful bacteria from human waste contaminate water and make it unfit to drink or swim in. In other instances such as toxic substances from industrial processes, it may take years to build up in the environment and food chain before their effects are fully recognized. The rapid increase in population and unplanned use of surface and groundwater is a major reason of water pollution. In 2003, the wastewater generation was estimated around 1500 km^3/year, which was more than six times of water in rivers and lakes of the world [26]. Around 2.4 billion people (30%) of the population are living without sanitation facility [27], because the water pollution is connected with population and industrial development [27]. The lack of facility for sanitation is also source of water contamination, which is one of the major reasons of pollution. The developing countries are the major contributor of water pollution. These countries are discharging 90% of sewage water directly in surface water bodies without any proper treatment [27]. About 730 million tons of waste is discharged into the water [28]. The share of industry in waste is 300–400 tons/year. It is estimated that due to degradation of resources of freshwater around 30% of global diversity has been lost so far [29]. Excess use of fungicide, insecticide, and herbicides are equally responsible for water resource pollution, contain toxic and carcinogenic substances, and heavily affect human and wildlife. Pesticides are also equally responsible for biodiversity and destroying insects and weed, which have negative impacts further on the food chain. About 120 million tons of nitrogen converts from the atmosphere in form of reactive nitrogen each year and around 2/3 of its goes to water bodies [30]. The fertilizer is very important source of phosphorus, nitrogen, and other macro- and micro-nutrients for agriculture, and as per reports use of fertilizer was 90 million ton/year in 2000 and may increase dramatically up to 150 million/year to increase the production to feed the population by 2050. The use of phosphorus and nitrogen can also increase up to 150–180% by 2050 [15]. Eutrophication in lakes of European countries such as Rhine, Danube, and Elbe Rivers is becoming a serious issue during 1970s–1980s. Water quality is failing to meet the standards and cause of oxygen-level depletion. Yet in 2009, around 20% of European lakes suffered from nutrient enrichment. Not this but nitrate concentration is increasing in groundwater resources [31]. Water is a basic need of all industrial sectors in different operational processes. It is estimated that around 5–20% of used water in industries turns into the polluted water and enters into the surface water bodies, rivers, canals, and pollute freshwater, as well as sea. Around 7% of untreated industrial wastewater is directly discharged into water and ultimately affects freshwater resources [32]. The largest portion of waste producer is countries from high-level income group. The share of these countries is around 70 percent in wastewater treatment; dramatically it is 38 percent share of countries from upper-middle income, while there is 28 percent share in wastewater treatment by lower-middle income countries and the lowest share is from low-income countries, which is only 8 percent of generated wastewater by industrial sectors [33].

The policies and seriousness of the countries are toward the main threat in the future; the share of the European countries in treatment is 71% from industrial and domestic/municipal sectors. Other side of around 20 percent shares of countries is from Latin America, while 51 percent the largest share is in treatment by Middle East and North Africa as shown in **Table 2**. Due to lack of funds by African countries, it is very difficult to treat the industrial and municipal wastewater generated by industries and domestic. It is reported that no data of wastewater generation, discharge, and treatment of 32 countries out of 48 from African countries are available in government- or water-related agencies [33]. Wastewater is the biggest threat for toxic and polluted wastewater from different industries, textile, sugarcane, pharmaceuticals, paper, food and some other industries. Around 2.2 million people die per year only by diarrhea diseases [38]. Around 10.4 million deaths of infants of about 17% are caused by the disease only. Among 80% of diarrhea cases, the reason behind is only lack of safe drinking water, poor facilities of sanitation, and unhygienic conditions. Use of nitrogen in different crops in agriculture sector is around 120 million tons converting from atmosphere into reactive nitrogen comprising of compounds every year [30]. About 2/3 of it goes into lands and in water sector bodies as well as in coastal zones. Approximately 150 million tons project the use of fertilizer by 2050, as compared to 90 million tons in year 2000 [39]. Due to the use of phosphorus and nitrogen, effluent will jump up to 150 and 180 percent by 2050, respectively [15]. The chemicals used in agriculture have enhanced up to 02 million tons each year, and around share of fungicides is 17.5 percent, insecticides 29.5 percent, and herbicides 47.5 percent and other's share is 5.5 percent. The economic growth and unplanned growth of population are also of great threats and main responsible for deterioration of quality of the water. Almost rivers of Asia, Latin America, and Africa are suffering from severe pollution level. Groundwater status in Europe was not perfect. Some stations of Europe were detected above level of nitrate in drinking water, recorded by the monitoring stations. As per reports, rivers around 30 percent and lakes 40 percent recorded eutrophic during the period of 2008–2011.The agriculture cultivation by wastewater is directly related to health issues, especially food (vegetables) in urban areas. About 10 percent of the population of world depends only on the food cultivated by polluted wastewater [40]. Around 26 percent productions of vegetables directly use wastewater in Pakistan; this situation prevails in all regions of the world [33]. Around 80 percent of total generated wastewater is directly discharged without fulfilling the requirements of regulatory bodies [41]. The treated wastewater usage is the largest and vital source for agriculture and other sectors. The use of safe-treated wastewater is gradually increasing its availability, especially in many Arab and other developing countries. It is reported that 71 percent of wastewater was effectively treated in Arab countries during 2013 and 21 percent out of 71 percent was utilized in agriculture sector (irrigation) and recharge of groundwater [41]. The treated wastewater use has become very important for water sector. At present, the market of wastewater is more than 700 billion dollars per year and increasing dramatically. In 2012, around 824,000 deaths occurred due to lack of the facilities of sanitation, drinking water, and proper hand washing in low- and middle-income countries. The

Counties	Treated municipal wastewater km^3 (Year)	Produced municipal wastewater km^3 (Year)	Direct use of treated municipal waste water in irrigation Km^3 (Year)
Japan	11.56 (2011)	16.93 (2011)	0.012 (2009)
Mexico	3.08 (2011)	7.46 (2011)	0.40 (2010)
China	26.61 (2009)	37.98 (2010)	0.48 (2008)
United States	40.89 (2008)	60.40 (2008)	0.33 (2004)
Germany	5.18 (2007)	5.30 (2007)	
Brazil	2.51 (2009)	9.73 (2009)	0.008 (2008)
Republic Korea	6.58 (2011)	7.84 (2011)	
India	4.42 (2011)	15.44 (2011)	
Egypt	3.71 (2012)	7.08 (2012)	
Canada	3.55 (2009)	6.61 (2009)	
Malaysia	2.60 (2009)	4.22 (2009)	
Thailand	1.17 (2012)	5.11 (2012)	
Turkey	2.72 (2010)	3.58 (2010)	
United Kingdom	4.05 (2011)	4.09 (2011)	
Italy	3.9 (2007)	3.93 (2007)	
France	3.77 (2008)	3.79 (2008)	
South Africa	1.92 (2009)	3.54 (2009)	
Iran	0.89 (2012)	3.55 (2010)	
Pakistan	0.55 (2011)	3.06 (2011)	
Spain	3.16 (2004)	3.18 (2004)	
Argentine	0.29 (2000)	2.46 (2010)	
Columbia	0.60 (2010)	2.40 (2010)	
Poland	1.36 (2011)	2.27 (2011)	
Vietnam	0.20 (2012)	1.98 (2012)	
Netherlands	1.88 (2010)	1.93 (2010)	
Australia	2.00 (2013)	1.83 (2007)	
Saudi Arabia	1.06 (2010)	1.55 (2010)	
Peru	0.10 (2012)	1.00 (2011)	

Table 2.
Municipal wastewater production, collection, and treatment in countries with the largest urban populations [33–37].

contaminated drinking water was one of the biggest sources of other causalities as well [42]. The water-related diseases are widely source of diarrhea in low middle- and low-income countries. Around 275,000 square kilometers of land was irrigated throughout world [34]. For example, the percentage of wastewater release without treatment has been estimated at 77% for Thailand (2012), 82% for Pakistan (2011), 84% for Armenia (2011), and 81% for Vietnam. Improving the efficiency of waste-water management would contribute to the achievement of the 2030 Agenda for

Sustainable Development in the region [41]. Approximately 15 percent of irrigated land can be cultivated by treated wastewater because around 330 square kilometers of municipal wastewater are generated per year in world and can be used for irrigation/cultivation on 40 million hectares (8000 m^3/hectares) [43]. Irrigation by wastewater or diluted wastewater is still fragmentary around 5–20 million hectares, almost lion share of China [44]. It is between the ranges of 2 and 7% of globally total cultivated area. Around 500 children's death is due to diarrhea-related diseases in India, which is contaminated water causing diseases. About 106 billion dollars or 5.2 percent of GDP is losing India from lack of sanitation and other facilities. On other side Israel has dramatically reduced its agricultural water budget from 70 percent to 55 percent and reduced to 15 percent in twenty years (1985–2005).

6. Conclusion

Economic development cannot declare whole reason for the detoriation of water quality as well as no solution for all problems. Some other steps are also very essential such as watershed management, which is very important for the water quality, water availability increase, and recharge of groundwater and helpful for reduction of other impacts. It is essential for managing irrigation water efficiency and protects the water quality. Allocation of water for agriculture is very hard/core issue and accurate data can play very important role for water conservation/saving and due to this many lives can be saved, which leads to cost reduction and minimizes other impacts on water resources/reservoirs. It is recommended to be a creative mind for the water management system such as municipal, industrial, and irrigation. The policy should be formulated on every country's basis, and should focus on less water requirement crops and latest techniques of irrigation system, such as micro-irrigation, and drip and sprinkler irrigation system, should be introduced/implemented to replace the traditional irrigation system. Crops with high yield on the lowest water use should be introduced to feed the population of around 9 billion by 2050, because population increase by 2050 will severely affect water resources and may reduce the availability of water. For policy making, it is necessary for data availability of wastewater generation, and its discharge without treatment or treated, without complete or proper information, is very difficult to make action plan and protection of environment resources from wastewater pollution. Given the importance of water we need to save water and change our life style to minimize use of water. It is difficult to imagine life on the earth without water. We should adopt agriculture mechanization, high-yielding seeds, and crops to be cultivated on less water, by replacing new irrigation techniques including drip irrigation and sprinkle irrigation, by water savings to cope with the growing population by 2050. With improving irrigation efficiency, the usage of freshwater can be reduced. It is recommended to monitor the entire water management system including agriculture to prevent from a severe crisis of water management and for that, we should be more creative than we ever were. For efficient usage of water, there is a need for the development of Global Monitoring Information System for water, which will provide us the necessary information on water management and progress monitoring of desirable targets. Development of this system will provide us the wide spectrum of information from different levels, like from local to national and ultimately the global level. Hurdles to research, technology, and innovation applications are due to the lack of financial resources for improvement of existing technologies in developing countries but also the lack of awareness for promoting the research and innovation in the countries that are already developed to shift to the new and efficient technologies for large-scale applications. The high costs of high-end technologies hamper their widespread application, especially in developing countries. It is because of the quick depletion of the available resources of water and its impact on climate change, conventional irrigation systems must be revised because it cannot sustain in many areas around the globe. In brief, the demand for water will increase by 2050 but the availability of water will be reduced. Water resources will reduce. Pollution will further reduce the amount of clean freshwater.

Conflict of interest

The authors declare no conflict of interest.

Author details

Aijaz Panhwar[1*], Rashid Abro[2], Aftab Kandhro[3], Abdul Rauf Khaskheli[4], Nusrat Jalbani[1], Khadim Ali Gishkori[5], Atta Muhammad Mahar[1] and Sofia Qaisar[2]

1 Pakistan Council of Scientific and Industrial Research Laboratories Complex, Karachi, Sindh, Pakistan

2 Department of Chemical Engineering, Dawood University of Engineering and Technology, Karachi, Sindh, Pakistan

3 M. A. Kazi Institute of Chemistry, University of Sindh, Jamshoro, Sindh, Pakistan

4 Shaheed Mohtarma Benazir Bhutto Medical University, Larkana, Sindh, Pakistan

5 Pakistan Standards and Quality Control Authority, Karachi, Sindh, Pakistan

*Address all correspondence to: aijazap@yahoo.com

References

[1] United Nation International Children Emergency Fund/World Health Organization. Progress on Drinking Water, Sanitation and Hygiene: Update and SDG Baselines. Geneva, Switzerland: WHO; 2017

[2] Food Agriculture Organization. Coping with Water Scarcity: Q & a with Food Agriculture Organization Director-General Dr Jacques Diouf. Newsroom: Food Agriculture Organization; 2007

[3] Burek P, Satoh Y, Fischer G, Kahil MT, Scherzer A, Tramberend S, et al. Water Futures and Solution: Fast Track Initiative (Final Report). IIASA Working Paper. Laxenburg, Austria: International Institute for Applied Systems Analysis; 2016

[4] Food Agriculture Organization. The State of the World's Land and Water Resources for Food and Agriculture: Managing Systems at Risk. Rome/London: FAO/Earthscan; 2011

[5] United Nation World Water Assessment Programme. The United Nations World Water Development Report. Geneva, Switzerland: United Nation World Water Assessment Programme; 2018

[6] Mekonnen MM, Hoekstra AY. Four billion people facing severe water scarcity. Science Advances. 2016;**2**(3):1-6

[7] Donnelly K, Cooley H. Water Use Trends in the United States. Oakland, Calif: Pac. Inst; 2015

[8] NGWA. Facts about Global, Groundwater Usage Compiled by NGWA. USA: NGWA; 2016

[9] CGWB. Ground Water Brochure, Medak District, Andhra Pradesh, India. 2013

[10] Where is Earth's water. United States Geological Survey. Available from: https://web.archive.org/web/20131214091601/http://ga.water.usgs.gov/edu/earthwherewater

[11] Wada Y, Flörke M, Hanasaki N, Eisner S, Fischer G, Tramberend S, et al. Modelling global water use for the 21st century: The water futures and solutions (WFaS) initiative and its approaches. Geoscientific Model Development. 2016;**9**:175-222

[12] United Nations Department of Economic and Social Affairs. World Population Prospects: Key Findings and Advance Tables – The 2017 Revision. Working Paper No. ESA/P/WP/248. New York: UNDESA, Population Division; 2017

[13] Organisation for Economic Co-operation and Development. Organisation for Economic Co-Operation and Development, Environmental Outlook to 2050: The Consequences of Inaction. Paris: OECD Publishing; 2012

[14] Alexandratos N, Bruinsma J. World Agriculture Towards 2030/2050: The 2012 Revision. ESA Working Paper No. 12-03. Rome: Food and Agriculture Organization of the United Nations (FAO); 2012

[15] Organisation for Economic Co-operation and Development. OECD Data: GDP Long-Term Forecast (Indicator). Paris, France: Organisation for Economic Co-operation and Development; 2017

[16] United Nation World Water Development Report. Nature Based Solution for Water. 2018

[17] Gleick PH, Palaniappan M. Peak water limits to freshwater

withdrawal and use. Proceedings of the National Academy of Sciences. 2010;**107**(25):11155-11162

[18] Hoekstra AY, Mekonnen MM. The water footprint of humanity. Proceedings of the National Academy of Sciences. 2012;**109**(9):3232-3237

[19] Food and Agriculture Organization of the United Nations/International Fund for Agricultural Development/United Nations Children's Fund/World Food Programme/World Health Organization. The State of Food Security and Nutrition in the World 2017: Building Resilience for Peace and Food Security. Rome: World Health Organization; 2017

[20] Park K. Preventive and Social Medicine. 25th ed. Prem Nagar, Jabalpur, India: M/s Banarsidus Bhanot; 2007

[21] Veldkamp TIE et al. Water scarcity hotspots travel downstream due to human interventions in the 20th and 21st century. Nature Communications. 2017;**8**:15697

[22] Scanlon BR et al. Global evaluation of new GRACE mascon products for hydrologic applications. Water Resources Research. 2016;**52**:9412-9429

[23] Qadir M, Sharma BR, Bruggeman A, Choukr-Allah R, Karajeh F. Non-conventional water resources and opportunities for water augmentation to achieve food security in water scarce countries. Agricultural Water Management. 2007a;**87**:2-22

[24] Pimentel D, Bailey O, Kim P, Mullaney E, Calabrese J, Walman L, et al. Will limits of the Earth's resources control human numbers. Environment, Development and Sustainability. 1999;**1**:19-39

[25] Rijsberman FR. Water scarcity: Fact or fiction? Agricultural Water Management. 2006;**80**:5-22

[26] United Nation World Water Assessment Program. Water for People, Water for Life: 3rd World Water Forum in Kyoto, Japan. Geneva, Switzerland: United Nation World Water Assessment Program; 2003

[27] United Nation International Children Emergency Fund/World Health Organization. Progress on Sanitation and Drinking Water 2015, Update and MDG Assessment. Geneva, Switzerland, 2016: JM Program; 2015

[28] The United Nations world water development report. Wastewater: The Untapped Resource. New York, United States: The United Nations World Water Development Report (United Nations Educational, Scientific and Cultural Organization; 2017

[29] United Nations, Water. Wastewater Management-a UN-Water Analytical Brief 1-52. Geneva, Switzerland: World Meteorological Organization; 2015

[30] Rockström J, Steffen W, Noone K, Persson Å, Chapin FS III, Lambin E, et al. Planetary boundaries: Exploring the safe operating space for humanity. Ecology and Society. 2009;**14**(2):32

[31] Ralf I, Kari A, Dietrich B, Benjamin B, Wera L, Evelyn L, et al. European Assessment of Eutrophication Abatement Measures across Land-Based Sources, Inland, Coastal and Marine Waters. European Environment Agency, European Topic Centre on Inland, Coastal and Marine Waters. ETC/ICM Technical Report 2/2016. Germany: Helmholtz Centre for Environmental Research GmbH-UFZ; 2016

[32] United Nations World Water Assessment Programme. The United Nations World Water Development Report. Geneva, Switzerland: United

Nations World Water Assessment Programme; 2009

[33] Sato T, Qadir M, Yamamoto S, Endo T, Zahoor M. Global, regional, and country level need for data on wastewater generation, treatment, and use. Agricultural Water Management. 2013;**130**:1-13

[34] AQUASTAT. Area Equipped for Irrigation. Infographic. Rome: Food and Agricultural Organization of the United Nations; 2014

[35] Water Resource Department of Land and Water Bureau. Water Resources in Japan. Japan: Ministry of Land Infrastructure, Transport, and Tourism Japan; 2012

[36] Environment Canada. Municipal Water Use Report: Municipal Water Use. 2006 Statistics. Gatineau: Environment Canada; 2010

[37] Jiménez B, Asano T. Water Reuse: An International Survey of Current Practice, Issues and Needs. London: IWA; 2008

[38] United Nation Development Program. Human Development Report, 2006. Beyond Scarcity: Power, Poverty and the Global Water Crisis. New York, United States: United Nation Development Program; 2006. p. 440

[39] Winiwarter W, Erisman JW, Galloway JN, Klimont Z, Sutton M. Estimating environmental loads of reactive nitrogen in the 21st century. Climatic Change. 2013;**120**:889-901

[40] World Health Organization/Food Agriculture Organization. Guidelines for the Safe Use of Wastewater, Excreta and Grey Water in Agriculture and Aquaculture. third ed. Vol. 1-4. Geneva: World Health Organization; 2006

[41] United Nations World Water Assessment Programme. The United Nations World Water Development ReportWastewater: The Untapped Resource. Paris: United Nations World Water Assessment Programme; 2017

[42] World Health Organization. Preventing Diarrhoea through Better Water, Sanitation and Hygiene: Exposures and Impacts in Low and Middle Income Countries. Geneva: Switzerland; 2014

[43] Mateo-Sagasta J, Raschid-Sally L, Thebo A. Global wastewater and sludge production, treatment and use (Chapter-02). In: Wastewater: Economic Asset in Urbanizing World. Netherlands: Springer; 2015

[44] Drechsel P, Evans AEV. Wastewater use in irrigated agriculture. Irrigation and Drainage Systems. 2010;**2010**(24): 1-3. DOI: 10.1007/s10795-010-9095-5